Novel Thermal and Non-thermal Technologies towards Sustainability and Microbiological Food Safety and Quality

Novel Thermal and Non-thermal Technologies towards Sustainability and Microbiological Food Safety and Quality

Guest Editors

Sotirios Oikonomou
Anastasia Kyriakoudi

MDPI

Basel • Beijing • Wuhan • Barcelona • Belgrade • Novi Sad • Cluj • Manchester

Guest Editors
Sotirios Oikonomou
School of Applied Sciences
University of the West of England
Bristol
UK

Anastasia Kyriakoudi
Department of Food Science
and Technology
Aristotle University of Thessaloniki
Thessaloniki
Greece

Editorial Office
MDPI AG
Grosspeteranlage 5
4052 Basel, Switzerland

This is a reprint of the Special Issue, published open access by the journal *Applied Sciences* (ISSN 2076-3417), freely accessible at: https://www.mdpi.com/journal/applsci/special_issues/0T594671X5.

For citation purposes, cite each article independently as indicated on the article page online and as indicated below:

Lastname, A.A.; Lastname, B.B. Article Title. *Journal Name* **Year**, *Volume Number*, Page Range.

ISBN 978-3-7258-7178-0 (Hbk)
ISBN 978-3-7258-7179-7 (PDF)
https://doi.org/10.3390/books978-3-7258-7179-7

Contents

About the Editors

Sotirios Oikonomou

Sotirios Oikonomou is a Post-doctoral Researcher in the School of Applied Sciences at the University of the West of England, Bristol, where he works within the Food Science area. His research focuses on food microbiology and safety, with particular emphasis on novel decontamination technologies such as cold atmospheric plasma, the development of electrochemical (bio)sensors for assessing food quality and authenticity, and the design of 3D printed foods. He is also interested in the behaviour of foodborne pathogens and beneficial microorganisms in complex matrices, including agri-food waste. Dr. Oikonomou possesses substantial breadth and depth of specialist knowledge in food microbiology, food safety and quality, food chemistry, electrochemical biosensors, and biotechnology applications. His research aims to develop novel, sustainable, and safe food products. Finally, his editorial roles, including service on the Early Career Editorial panel for the MDPI journal *Applied Sciences*, demonstrate recognised expertise and communication skills within the research community, with a view to building solid relationships for future large-scale collaborations and expanding his network.

Anastasia Kyriakoudi

Anastasia Kyriakoudi is a Chemist with a master's degree in "Food Chemistry and Technology" from the School of Chemistry (AUTH). She holds a PhD in "Plant Extracts Rich in Isoprenoid Compounds of Interest in Food Bioactive Ingredients Industry". In 2015, she carried out post-doctoral studies at the Italian National Agency for New Technologies, Energy and Sustainable Economic Development, ENEA (Roma, Italy), on geographical traceability issues, and in 2016, she was awarded an IKY–Siemens fellowship in the field of "Nanotechnology and Applications" for post-doctoral studies in Greece. She is currently a Post-doctoral Researcher at the Laboratory of Food Chemistry and Biochemistry of the School of Agriculture (AUTH). Her research interests include the valorisation of by-products/wastes of the food industry into high-added value products (e.g., natural antioxidants, colorants), the extraction of secondary metabolites from natural products using green alternative solvents (e.g., NADES) and techniques (e.g., ultrasound- and microwave-assisted extraction), the characterization of extracts using several analytical techniques (e.g., UV-Vis, HPLC-DAD, LC-MS, GC-FID, GC-MS, FT-IR), the examination of in vitro antioxidant activity, the bioaccessibility/bioavailability of bioactive food ingredients, as well as their encapsulation (e.g., inclusion complexation, freeze-drying, nano spray-drying), and the optimization of food-related processes. As a master's student, PhD candidate, and Post-doctoral Researcher, she has also participated in several national and European projects.

applied sciences

MDPI

Editorial

Novel Thermal and Non-Thermal Technologies Towards Sustainability and Microbiological Food Safety and Quality

Sotirios I. Ekonomou

Centre for Sustainable Agri-Food and Environment (SAFE), Faculty of Health and Applied Sciences, University of the West of England, Bristol BS16 1QY, UK; sotirios.oikonomou@uwe.ac.uk; Tel.: +44-1179656261

1. Introduction

Ensuring food safety and quality while addressing sustainability challenges remains a cornerstone of modern food systems. Conventional preservation and processing methods, while effective, are increasingly being scrutinised for their energy demands, nutrient degradation, and limited ability to align with consumer expectations for minimally processed, clean-label, and environmentally friendly foods [1]. Conventional food preservation methods use high temperatures to inhibit foodborne pathogens and reduce the microbial load [2], thereby extending the shelf-life of food. However, this process can lead to the loss of essential temperature-sensitive nutrients, alter the texture of food products, and change their sensory characteristics. Pasteurisation, sterilisation, cooking, and drying are currently the main technological approaches for food preservation, which have replaced traditional preservation techniques [3]. These conventional heating techniques primarily depend on producing heat externally to the item being heated, either through burning fuels or using an electric resistive heater. This results in high energy consumption, significant financial costs, and a substantial environmental footprint. Thus, the collaboration between processors and academic institutions to meet consumer demands for high-quality, sustainably produced food, while addressing rising economic standards, has led to the emergence of new technological approaches in food processing. At the same time, the global burden of foodborne illness and spoilage underscores the need for robust technologies capable of achieving reliable microbial inactivation.

Non-thermal technologies are food processing methods that do not rely on high temperatures. They are emerging in the food industry due to their potential to enhance food safety, quality, and sustainability [4]. Some of the most advantageous non-thermal techniques that have emerged, showing notable spoilage inhibition and retention of the sensory properties of foods, are high hydrostatic pressure (HHP), ultrasound, UV radiation (UV-C), ozone, pulsed electric field, and, most recently, cold plasma (CP). Non-thermal technologies can preserve the quality of food and extend its shelf-life by deactivating spoilage microorganisms in various food products, including seafood [5], pork [6], chicken [7], and fresh fruit [8], among others.

This Special Issue, "Novel Thermal and Non-thermal Technologies towards Sustainability and Microbiological Food Safety and Quality", was launched to address these challenges and to showcase advances in both fundamental research and applied studies. The contributions collectively explore how emerging processing strategies can support safe, high-quality, and sustainable food production, offering insights into microbial inactivation, product functionality, and environmental impact.

2. Overview of Contributions

This Special Issue compiles eight original articles and one review that advance knowledge on novel thermal, non-thermal, and hybrid technologies for safe, high-quality, and sustainable food processing. High-pressure processing (HPP) was extensively investigated as a non-thermal pasteurisation alternative. Shad et al. [9] compared HPP (600 MPa, 4 min) with conventional heat treatment (82 °C, 5 min) for pesto sauces containing 34% and 54% oil and inoculated with *Listeria monocytogenes* and *Salmonella Typhimurium*. They observed >4 log pathogen reduction, with no significant effect of oil content on inactivation, and superior preservation of lipid oxidation indices and sensory quality during refrigerated storage.

Hybrid drying methods combining thermal and non-thermal energy inputs were also reported by Mierzwa and Musielak [10]. They showed that microwave-assisted rotary drying of carrots increased drying rates by 112% and microwave–ultrasound combinations by 140%, while reducing specific energy consumption by approximately 30%, with some impact on colour parameters. Similarly, Gondek et al. [11] demonstrated that microwave–convection drying of Moldavian dragonhead leaves drastically shortened the drying time and resulted in higher retention of polyphenols, chlorophyll, and antioxidant activity compared with conventional convection drying—an approach that valorises agricultural by-products.

Two papers addressed the critical link between storage temperature, microbial growth, and sensory acceptance. Frangopoulos et al. [12] modelled aerobic plate count kinetics in pasteurised orange juice under accelerated storage (10–40 °C), reporting shelf-life extension at lower temperatures and rapid spoilage at elevated ones. In a companion sensory study, the same authors [13] found that consumer rejection typically preceded microbiological unsafety by 3 days, reinforcing the need to define shelf-life based on both sensory and safety endpoints.

At the ingredient level, Tsioptsias et al. [14] characterised the thermal behaviour and infrared spectral changes of citric acid monohydrate during dehydration, recrystallisation, and mild heating using Differential Scanning Calorimetry, Thermogravimetric Analysis, and computational modelling, offering insights for formulation stability and potential effects on microbial control in acidified products.

In addition to processing studies, this Special Issue also included contributions that focused on analytical innovations for food quality monitoring. Ekonomou et al. [15] developed a disposable amperometric biosensor based on a screen-printed carbon electrode modified with Meldola's Blue and coated with glycerol dehydrogenase/NAD^+, designed for rapid glycerol determination in wine. The biosensor exhibited good linearity in the 1–7.5 mM range, accurate agreement with a reference spectrophotometric method, and required only a small sample volume. Such tools can facilitate on-site monitoring of fermentation and product quality, supporting early decision-making and reducing waste.

Finally, Russell et al. [16] provided a narrative review on the potential of molecular hydrogen (H_2) applications across the farm-to-fork continuum, including postharvest treatment, storage, and processing, where it may act as an antioxidant, delay senescence, and reduce microbial spoilage, thereby contributing to food system sustainability.

Collectively, these studies highlight that novel thermal and non-thermal technologies can deliver effective microbial inactivation while maintaining nutritional and sensory quality. They also stress the importance of optimising process parameters, understanding storage effects, developing real-time monitoring tools, and integrating sustainability considerations, paving the way for wider adoption of such approaches in industry.

3. Conclusions

This Special Issue brings together a diverse set of studies that collectively highlight the potential of novel thermal and non-thermal food processing technologies to enhance microbiological safety, protect food quality, and contribute to sustainability. The breadth of work illustrates the importance of multidisciplinary collaboration among microbiologists, engineers, food technologists, and sustainability scientists.

Funding: This research received no external funding.

Conflicts of Interest: The author declares no conflicts of interest.

References

1. Delgado-Pando, G.; Ekonomou, S.I.; Stratakos, A.C.; Pintado, T. Clean Label Alternatives in Meat Products. *Foods* **2021**, *10*, 1615. [CrossRef] [PubMed]
2. Ekonomou, S.I.; Boziaris, I.S. Fate of Osmotically Adapted and Biofilm Listeria Monocytogenes Cells after Exposure to Salt, Heat, and Liquid Smoke, Mimicking the Stresses Induced during the Processing of Hot Smoked Fish. *Food Microbiol.* **2024**, *117*, 104392. [CrossRef] [PubMed]
3. Pereira, R.N.; Vicente, A.A. Environmental Impact of Novel Thermal and Non-Thermal Technologies in Food Processing. *Food Res. Int.* **2010**, *43*, 1936–1943. [CrossRef]
4. Ekonomou, S.I.; Boziaris, I.S. Non-Thermal Methods for Ensuring the Microbiological Quality and Safety of Seafood. *Appl. Sci.* **2021**, *11*, 833. [CrossRef]
5. Ekonomou, S.I.; Bulut, S.; Karatzas, K.A.G.; Boziaris, I.S. Inactivation of Listeria Monocytogenes in Raw and Hot Smoked Trout Fillets by High Hydrostatic Pressure Processing Combined with Liquid Smoke and Freezing. *Innov. Food Sci. Emerg. Technol.* **2020**, *64*, 102427. [CrossRef]
6. Oliinychenko, Y.K.; Ekonomou, S.I.; Tiwari, B.K.; Stratakos, A.C. Assessing the Effects of Cold Atmospheric Plasma on the Natural Microbiota and Quality of Pork during Storage. *Foods* **2024**, *13*, 1015. [CrossRef] [PubMed]
7. Naseem, T.; Zahid, U.; Shahzad, A.; Hassan, S.A.; Abdi, G.; Aadil, R.M. Cold Plasma as a Frontier in Combating Foodborne Bacterial Pathogens in Ready-to-Eat (RTE) Foodstuff. *Appl. Food Res.* **2025**, *5*, 100842. [CrossRef]
8. Song, C.; Wang, J.; Wu, L.; Liu, J.; Liu, G.; Gong, D.; Zhang, W.; Wei, J.; Zhang, Z. Quality and Physiological Changes in Fresh-Cut Mango Fruit as Affected by Cold Plasma-Activated Water. *Postharvest Biol. Technol.* **2025**, *225*, 113524. [CrossRef]
9. Shad, E.; Raninen, K.; Podergina, S.; Chan, L.I.; Tong, K.P.; Hälikkä, H.; Huovinen, M.; Korhonen, J. Impact of High-Pressure Processing on Quality and Safety of High-Oil-Content Pesto Sauce: A Comparative Study with Thermal Processing. *Appl. Sci.* **2024**, *14*, 9425. [CrossRef]
10. Mierzwa, D.; Musielak, G. Microwave and Ultrasound Assisted Rotary Drying of Carrot: Analysis of Process Kinetics and Energy Intensity. *Appl. Sci.* **2024**, *14*, 10676. [CrossRef]
11. Gondek, E.; Kamińska-Dwórznicka, A.; Kocira, S.; Oniszczuk, T.; Bialik, M.; Stasiak, M. Convection and Microwave–Convection Drying of Moldavian Dragonhead (*Dracocephalum moldavica* L.) Leaves. *Appl. Sci.* **2024**, *14*, 11496. [CrossRef]
12. Frangopoulos, T.; Koliouskas, A.; Petridis, D. The Effect of Accelerated Storage Temperature Conditions on the Shelf Life of Pasteurized Orange Juice Based on Microbiological, Physicochemical, and Color Attributes. *Appl. Sci.* **2024**, *14*, 10870. [CrossRef]
13. Frangopoulos, T.; Koliouskas, A.; Petridis, D. Sensory Shelf Life of Pasteurized Orange Juice Stored Under Different Temperature Levels Using Inverse Time Sampling and a Balanced Incomplete Block Design. *Appl. Sci.* **2025**, *15*, 1809. [CrossRef]
14. Tsioptsias, C.; Panagiotou, A.; Mitlianga, P. Thermal Behavior and Infrared Absorbance Bands of Citric Acid. *Appl. Sci.* **2024**, *14*, 8406. [CrossRef]
15. Ekonomou, S.I.; Crew, A.; Doran, O.; Hart, J.P. Development of a Disposable, Amperometric Glycerol Biosensor Based on a Screen-Printed Carbon Electrode, Modified with the Electrocatalyst Meldolas Blue, Coated with Glycerol Dehydrogenase and NAD+: Application to the Analysis of Wine Quality. *Appl. Sci.* **2024**, *14*, 6118. [CrossRef]
16. Russell, G.; Nenov, A.; Hancock, J.T. How Hydrogen (H2) Can Support Food Security: From Farm to Fork. *Appl. Sci.* **2024**, *14*, 2877. [CrossRef]

applied sciences

MDPI

Article

Development of a Disposable, Amperometric Glycerol Biosensor Based on a Screen-Printed Carbon Electrode, Modified with the Electrocatalyst Meldolas Blue, Coated with Glycerol Dehydrogenase and NAD^+: Application to the Analysis of Wine Quality

Sotirios I. Ekonomou, Adrian Crew, Olena Doran and John P. Hart *

Centre for Research in Biosciences, School of Applied Sciences, College for Health, Science and Society, University of the West of England, Coldharbour Ln, Bristol BS16 1QY, UK; sotirios.oikonomou@uwe.ac.uk (S.I.E.); adrian.crew@uwe.ac.uk (A.C.); olena.doran@uwe.ac.uk (O.D.)
* Correspondence: john.hart@uwe.ac.uk

Featured Application: Measurement of glycerol content for wine safety and quality.

Abstract: This paper describes the design and development of a novel electrochemical biosensor for measuring glycerol in wine. Our initial detailed studies were aimed at deducing the optimum conditions for biosensor operation by conducting hydrodynamic voltammetric and amperometric studies. The resulting voltammograms revealed a maximum electrocatalytic current at 0.0 V vs. Ag/AgCl, which we used for all further studies. We also examined the effect of pH (8–10) on the amperometric responses of different glycerol concentrations over a range of 0.04 to 0.20 mM. Based on our findings, we propose that pH 9 would be suitable as the supporting electrolyte for further studies with the amperometric biosensor. The biosensor was constructed by immobilising 10 units of GLDH and 660 μg NAD^+ onto the MB-SPCE surface using glutaraldehyde (GLA) as a cross-linking agent. Calibration studies were performed with glycerol over the 1.0–7.5 mM concentration range. Chronoamperometry was the electrochemical technique chosen for this purpose as it is convenient and can be performed with only 100 μL of sample directly deposited onto the biosensor's surface. In the current study, we observed linear calibration plots with the above standard solutions using current measurements at a selection of sampling times along the chronoamperograms (30–340 s). We have evaluated the glycerol biosensor by carrying out an analysis of commercially available red wine. Overall, these findings will form a platform for the development of novel rapid technology for point-of-test evaluation of glycerol in the production and quality control of wine.

Keywords: food safety; food quality; wine; amperometric glycerol biosensor; hydrodynamic voltammetry; chronoamperometry; screen-printed carbon electrode; meldolas blue

1. Introduction

Glycerol is a biologically important compound that is both a naturally occurring triose sugar and a food additive (E 422) [1]. Therefore, its concentration in foods and beverages needs to be controlled for food quality and safety purposes. Consequently, reliable analytical methods for determining it in different and often challenging matrices are urgently needed.

One of the most powerful analytical techniques used in measuring glycerol in beverages is High-Performance Liquid Chromatography (HPLC); some examples of analytical approaches developed for beverages are discussed below. Some include the possible chemical constituents accompanying glycerol in the beverage matrix to illustrate the important analytical challenges [2]. An HPLC method involving a cation ion exchange column, preceded by a guard column, in conjunction with a refractive index detector was described by Li et al. [3], which proved successful for the measurement of glycerol in beers. This group reported [3] that controlling glycerol was important, as levels below or above the threshold could be detrimental to the product's flavour. The authors showed complete chromatographic separation of glycerol from ethyl alcohol, as well as earlier eluting unidentified compounds; levels of glycerol were between 0.925 and 1.502 gL^{-1}, and ethanol ranged from 3.0 to 5.0% (v/v) in a variety of commercial beers.

HPLC methods have also been investigated in the context of wine analysis. Lopez and Gomez [4] successfully developed an HPLC method for wine analysis that could measure glycerol, as well as other sugars and carboxylic acids. The authors determined that the glycerol content of white wine was 6.45 gL^{-1}. A similar study was also reported by Calull et al. [5]. Moro et al. [6] reported on an HPLC method that was applied to the determination of glycerol, glucose, and fructose in wines to ascertain whether the drinks were high- or standard-quality red and white wines. This assay was achieved using HPLC in combination with a modified polar-bonded NH_2–silica phase column and a refractive index detector. The authors indicated that the concentration of glycerol in high-standard red wine was 3.30–7.67 gL^{-1} and in white wine, the levels were 1.65–5.87 gL^{-1}.

An interesting method involving HPLC combined with an FTIR detection system was applied to the determination of glycerol in red wine [7]. It was shown that lactic acid co-eluted with glycerol on the column, but the on-line FTIR detection system was able to resolve the responses of the two compounds for quantification. The glycerol concentration was determined for seven samples of red wine, and the range was reported to be 7.88–9.29 gL^{-1}.

Linget et al. [8] have developed a reproducible and high-throughput technique for on-line clean-up, coupled with HPLC, for routine analysis of glycerol and other small molecules such as amino acids. An ion exchange analytical column was used in conjunction with an RI detector, and the concentration found in red wine was found to contain 6.9 gL^{-1}. In addition to glycerol, Linget et al. [8] simultaneously determined a variety of other species by the reported method. In red wine (Cabernet Sauvignon), 0.45 gL^{-1} of citric acid, 1.99 gL^{-1} of tartaric acid, 0.46 gL^{-1} of malic acid, 0.7 gL^{-1} of succinic acid, 1.91 gL^{-1} of lactic acid, and 0.51 gL^{-1} of acetic acid were present and determined. Glucose 0.15 gL^{-1}, fructose, and ethanol 122 gL^{-1} were also present but did not appear to interfere with the determination of glycerol. The HPLC-based methods described above have shown how glycerol may be determined effectively in a range of alcoholic drinks, even in the presence of a variety of other naturally occurring species.

However, using this sophisticated instrument has drawbacks. These include the high cost of the instrument, which requires special expertise to operate and maintain, and the latter is often time-consuming. The actual sample preparation and chromatographic analysis may also be time-consuming.

In contrast, electrochemical biosensors have been developed based on screen-printing technology [9,10]. These devices have the advantage that they can be mass-produced in a wide range of geometries at low cost, particularly when carbon materials are employed to construct the electrodes. We explored the possibility of developing a screen-printed biosensor incorporating glycerol dehydrogenase (GLDH) as the biorecognition element. This enzyme requires the cofactor NAD^+ for the oxidation of glycerol, which results in the formation of NADH. Consequently, it was considered possible to measure glycerol indirectly by monitoring the generation of NADH during the oxidation reaction.

Our group was the first to develop an electrochemical strategy for the measurement of NADH based on a Meldolas Blue-Screen-printed Carbon Electrode (MB-SPCE) [11]. Meldolas Blue acts as an electrocatalyst for NADH oxidation, and the resulting anodic current constitutes the analytical response [11,12]. The main advantage of this approach is that the amperometric detection of NADH can be performed at potentials of 0.0 V vs. Ag/AgCl, whereas potentials in excess of +0.5 V vs. Ag/AgCl are required to obtain the limiting oxidation current at plain carbon electrodes [13]. Consequently, the selectivity is expected to be greatly improved with the former, which is highly beneficial for the analysis of real samples that may contain electroactive interferences.

We have exploited the above approach, employing biosensors based on MB-SPCEs to measure a variety of target analytes in different sample types, and all of the following examples have employed this platform.

A lactate biosensor was fabricated by coating the surface of the screen-printed electrode surface with lactate dehydrogenase and the cofactor NAD^+. This device was successfully applied to the measurement of lactic acid in serum [14]. This analysis could be performed by simply depositing 20 μL aliquots of solution onto the biosensor strip for chronoamperometric measurements. The resulting data demonstrated the possibility of applying the lactate biosensor to biomedical studies. In the area of environmental monitoring, a disposable screen-printed ammonium ion biosensor was constructed by immobilising the enzymes glutamate dehydrogenase, 2-oxoglutarate, and NAD_+ onto the electrode surface and was successfully applied to river water analysis [15]. This biosensor was subsequently incorporated into a commercial hand-held instrument for the measurement of ammonium ions in sewage effluent [16].

Additionally, we explored the possibility of adapting the MB-SPCE platform for the fabrication of amperometric biosensors for applications in the food industry. An amperometric glutamate biosensor was constructed by immobilising glutamate dehydrogenase onto the electrode surface after it had been modified with the biopolymer chitosan [17]. This was successfully applied to the determination of glutamate in a stock cube following a simple dissolution step in buffer solution, to which was added NAD^+. In order to circumvent the need to add NAD^+ to the sample, an improved glutamate biosensor was subsequently developed. This involved incorporating GLDH and NAD^+ into a hydrogel formed by chitosan and carbon multi-walled nanotubes (MWNTs), which were deposited onto the electrode surface using a layer-by-layer approach [18]. This novel biosensor was successfully applied to the determination of glutamate in a stock cube.

In recent years, we have investigated the possibility of developing a biosensor for androstenone, as this compound, together with skatole, is responsible for an off-taint in pig meat known as boar taint [19]. We immobilised the enzyme hydroxysteroid dehydrogenase (HSD) onto the screen-printed electrode by an adsorption procedure. It was demonstrated that androstenone could be measured by directly inserting the biosensor into the adipose tissue layer of the meat sample and that the androstenone concentration may indicate boar taint above a designated threshold level.

The electrocatalytic biosensor platform described above has also been adapted for the development of a biosensor for use in the alcoholic beverage industry. The MB-SPCE surface was modified by the immobilisation of ethanol dehydrogenase, NAD^+, and trehalose, using cellulose acetate to encapsulate the reaction layer. The resulting biosensor was successfully applied to measuring ethanol in gin [14].

The aim of the present study was to investigate the possibility of developing a robust, disposable electrochemical glycerol biosensor based on the MB-SPCE transducer. This biosensor would have important attributes, including mass production at a low cost and the possibility of rapid measurements by simply depositing small aliquots of liquid samples directly onto the flat biosensor surface. The chronoamperometric technique is very suitable for this purpose and is both simple to perform and easy to interpret. This approach would alleviate the requirement for laboratory glassware and help to achieve the portability of an analytical biosensor system. This paper describes the steps involved in fabricating and optimising the biosensor, as well as its application to measuring glycerol in wine. This method provides a foundation for the development of innovative, rapid point-of-test technology and should have a considerable impact on food and beverage quality and safety assessment.

2. Materials and Methods

2.1. Instrumentation

All voltammetric and amperometric measurements were carried out with an Emstat Blue potentiostat (PalmSens, Houten, The Netherlands), which was interfaced to a PC for data acquisition via PS Trace Software version 5.10.

MB-SPCEs were supplied by Gwent Electronic Materials Ltd. (Newport, UK). The working electrode was fabricated using carbon-based ink with CoPC (C2030408P3), and the reference electrode was fabricated using Ag/AgCl ink (C2130809D5). The working electrode area (3×3 mm) was defined using electrical insulation tape.

All pH measurements were performed using a Testo 205 (Testo Limited, Alton, Hampshire, UK) pH meter. Solutions were stirred using a colour squid (IKA, Tunbridge Wells, UK) and warmed using a HAAKE P5 water bath (Thermo Scientific, Loughborough, UK).

Optical digital microscopy (OM) was conducted using the Keyence optical microscope (Keyence, Itasca, IL, USA). The High Dynamic Range (HDR) imaging function was utilised to capture multiple images at varying shutter speeds, resulting in an image with enhanced colour gradation and contrast for a more detailed observation. Additionally, the optical shadow effect mode enabled the visualisation of surface details on the sensors. All images were captured at $80\times$ magnification.

The characterisation of the biosensor surface was performed using an attenuated total reflection (ATR) FTIR spectrometer (Nicolet 50, Thermo Fisher Scientific, Loughborough, UK) with built-in ATR. An analysis was performed to confirm the presence of all compounds on the full glycerol biosensor surface and a sample of glutaraldehyde. Scans were performed across a range of 4000–400 cm^{-1}, averaging 128 scans at a 4 cm^{-1} resolution with a 1 cm^{-1} interval. Background absorption was subtracted before analysis.

2.2. Raw Materials, Chemicals, and Reagents

Red wine (Merlot) from California was purchased at Sainsbury's, a UK supermarket chain, and stored at 4 °C. GLDH was obtained from Merck (Darmstadt, Germany). All other chemicals were purchased from Sigma Aldrich (Dorset, UK). Deionised water was obtained from a Purite RO200 Stillplus HP System (Oxon, UK). Stock solutions of monosodium, disodium, and trisodium orthophosphate were prepared at a concentration of 0.2 M by dissolving the appropriate mass in deionised water and were then titrated to achieve the desired pH and diluted in the cell to achieve a working concentration of 0.1 M. Sodium chloride was prepared to a concentration of 1.0 M by dissolving the appropriate mass in deionised water and was diluted in the cell, giving a final concentration of 0.1 M. Ascorbic oxidase (AO; 1000 U mg^{-1}, 189,724) from cucumber (*Cucurbita* sp.) was purchased by Merck (Darmstadt, Germany).

2.3. Glycerol Assessment

The glycerol estimation was obtained using a commercial glycerol assay kit (Megazyme International Ireland, Wicklow, Ireland). The glycerol is phosphorylated in this assay using adenosine-5′-triphosphate (ATP). The formed adenosine-5′-diphosphate (ADP) is used to phosphorylate d-glucose, which is oxidised with nicotinamide adenine dinucleotide (NADH) formation. The amount of NADH was detected by measuring absorbance at 340 nm using a UV-vis spectrophotometer (GENESYS 150, Thermo Scientific, UK).

2.4. Biosensor Fabrication

For the fabrication of the GLDH biosensors, the surface of the MB-SPCE working electrodes was coated with 10 μL of a mixture containing 10 U of GLDH and 660 μg of NAD^+ in a 0.1 M PBS, pH 9, solution. This layer was allowed to dry overnight under a vacuum at −0.6 MPa, at 4 °C. Next, a 10 μL aliquot of glutaraldehyde (0.01% in phosphate buffer) was deposited on top of the enzyme/NAD^+ layer and again left overnight, under vacuum, to allow crosslinking to occur. These biosensors were refrigerated until they were ready for use for calibration purposes and wine analysis.

2.5. Hydrodynamic Voltammetry of NADH in Alkaline Solutions Using an MB-SPCE

To deduce the optimum oxidation potential of NADH required for the operation of the biosensor, hydrodynamic voltammetry was performed with the MB-SPCE from an initial potential of −0.3 V to a final potential of +0.3 V vs. Ag/AgCl, and the current measurements were recorded at 50 mV intervals. Aliquots (10 mL) of either 0.1 M pH 8.0, 9.0, or 10 phosphate buffered saline (PBS) containing 5 mM NADH were transferred to the electrochemical cell, and the solutions were stirred at 250 rpm at a temperature of 30 °C. A fresh MB-SPCE was used to record the hydrodynamic voltammogram of the three buffer solutions.

2.6. Effect of pH on Amperometric Responses of Glycerol Solutions Containing Free GLDH and NAD^+

Next, we examined the effect of pH (8–10) on the amperometric responses of different glycerol concentrations. All the solutions were stirred at 250 rpm and contained 10 units of GLDH and 5 mM NAD^+, and the solution temperature was controlled at 30 °C. A fresh MB-SPCE was used to record the amperograms for each pH over the 0.04–0.2 mM concentration range.

2.7. Calibration Procedure and Analysis of Wine

Calibration studies were performed with glycerol solutions, prepared in 0.1 M PBS pH 9, over the 1.0–7.5 mM concentration range. The electrochemical technique employed was chronoamperometry, using an applied potential of 0.0 V vs. Ag/AgCl, preceded by an incubation time of 3 min. Aliquots of 100 µL of standards and samples were directly deposited onto the glycerol biosensor surface, which was warmed to 30 °C by placement onto a thermostated surface. A fresh biosensor was used to measure each concentration of glycerol. In the current study, we constructed calibration plots with the above standard solutions by taking current measurements at a selection of sampling times along the chronoamperograms at 30, 100, 200, 300, and 340 s.

For the analysis of red wine, an initial dilution step was carried out in order to adjust the concentration of glycerol to be in the linear range. Wine dilutions were carried out using 0.1 M PBS, pH 9. In addition, a final treatment of the sample was carried out with the addition of AO to enzymatically remove any ascorbic acid present in the sample.

2.8. Statistical Analysis

All data acquired were expressed as mean ± standard deviation (SD), and the Excel Microsoft® Office 365 (ver. 16.48) was used to plot graphs. Data were analysed using the paired two sample means *t*-test with IBM® SPSS® statistics 26 software for macOS (SPSS Inc., Chicago, IL, USA) at a 5% level of significance.

3. Results

3.1. Fabrication and Operation of the Glycerol Biosensor

Figure 1 presents the steps involved in fabricating the glycerol biosensor. Initially, a mixture containing GLDH and NAD^+ in phosphate buffer was deposited on a base of MB-SPCE. This layer was allowed to dry overnight under vacuum at 4 °C. Next, a 10 µL aliquot of a solution containing GLDH in PBS was deposited onto the enzyme/NAD^+ layer and again left overnight under vacuum to allow crosslinking to occur. These biosensors were refrigerated until they were ready for use.

Figure 2 shows the sequence of reactions that occur during the operation of the amperometric glycerol biosensor. The biosensor's operation principle is that glycerol is enzymatically oxidised by GLDH in the presence of NAD^+, which results in the production of NADH and dihydroxyacetone. NADH then undergoes electrocatalytic oxidation at an MB-SPCE, and the resulting electrocatalytic oxidation current constitutes the analytical response. This response is directly proportional to the glycerol concentration. It should be emphasised that the advantage of exploiting the electrocatalytic oxidation reaction of

NADH is that the operating potential is much lower (0.0 V vs. Ag/AgCl) than is possible by direct oxidation at bare carbon electrodes. This aspect is discussed later.

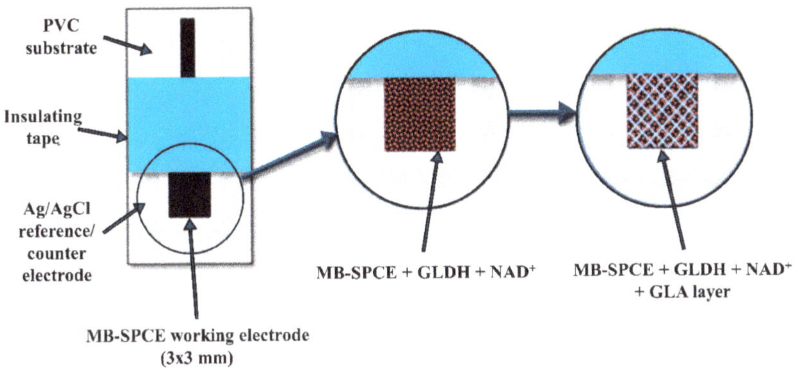

Figure 1. Schematic diagram of the biosensor fabrication involving the deposition of GLDH and NAD^+ followed by the application of the GLA cross-linking layer.

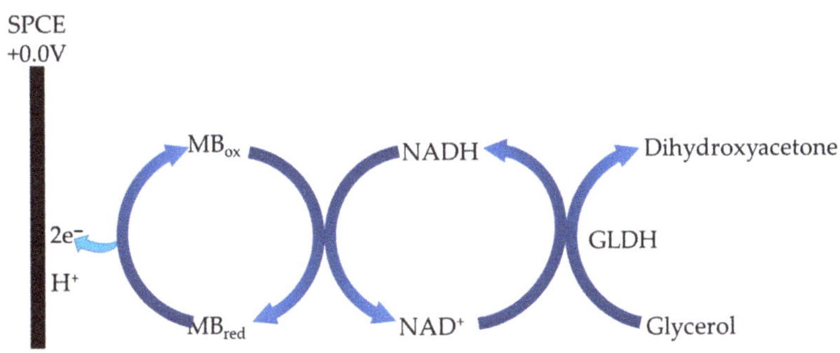

Figure 2. The sequence of reactions involved in the operation of the amperometric glycerol biosensor.

3.2. Surface Characterisation of the Biosensor

High-resolution digital microscopy was used to investigate the surface morphology of the glycerol biosensor during the fabrication process. Figure 3A–C displays three images captured during the process. Figure 3A shows the surface of the initial MB-SPCE, with clearly visible striations due to the screen design used in the screen printing process. Figure 3B depicts the image obtained after the deposition of the GLDH/NAD^+ mixture, showing an evenly dispersed layer over the surface, although the striations are still partially visible. Finally, Figure 3C represents the image obtained after the deposition of a GLA layer over the enzyme/cofactor layer, which had been allowed to dry and crosslink. This cross-linked layer encapsulates the underlying reaction layer, which results in strong immobilisation on the surface of MB-SPCE. This is evident from the latter image, where the upper layer is opaque and no striations are visible.

It was of interest to deduce the extent of cross-linking of glutaraldehyde with GLDH, so we next performed FTIR spectroscopic analysis on the biosensor surface. The FTIR spectrum obtained for glutaraldehyde alone (Figure 4) clearly shows a strong, sharp band at 1725 cm^{-1}, resulting from the C=O stretching mode. However, the spectrum obtained for the biosensor surface (Figure 4) does not exhibit this band, which indicates that a reaction of the C=O with the amino moiety of the enzyme has occurred. Therefore, the FTIR study indicates that complete cross-linking of glutaraldehyde with the enzyme has occurred, resulting in the formation of a Schiff base. Similar observations and conclusions have been reported previously by Lović et al. [20].

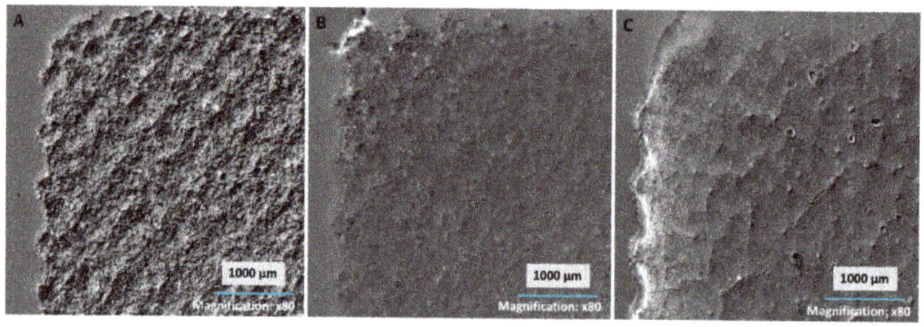

Figure 3. High-resolution (digital) microscopy images at 80× magnification; (**A**) the surface of the MB-SPCE; (**B**) the MB-SPCE with a deposited layer of GLDH/NAD^+; (**C**) the GLDH/NAD^+ layer cross-linked to the MB-SPCE surface with glutaraldehyde.

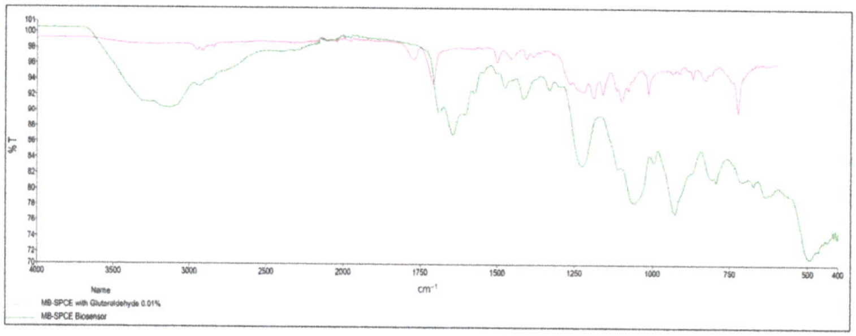

Figure 4. FTIR analysis of the glycerol MB-SPCE biosensor surface (green line) and glutaraldehyde alone (pink line).

3.3. Optimisation of the Biosensor Conditions

3.3.1. Hydrodynamic Voltammetry

Hydrodynamic voltammetric studies were performed with NADH under alkaline buffer conditions (pH 8–10) using the MB-SPCEs only. The resulting voltammograms exhibited a maximum electrocatalytic current at 0.0 V vs. Ag/AgCl (Figure 5). Therefore, this potential was used for all further studies. The electrocatalytic oxidation reaction is illustrated in Figure 2. This shows that 2 electrons per NADH molecule are transferred to the oxidised form of Meldolas Blue (MB_{ox}) to produce the reduced form (MB_{red}), which is subsequently transferred to the SPCE, generating the response current.

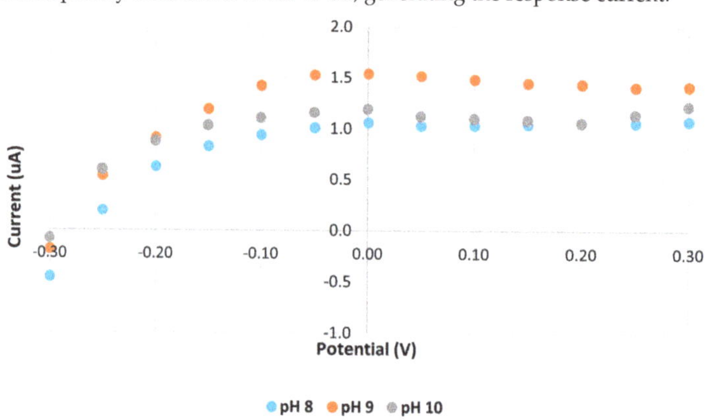

Figure 5. Hydrodynamic voltammograms for NADH were obtained using an MB-SPCE in buffer solutions of pH 8, 9, and 10.

3.3.2. Effect of pH on Amperometric Responses of Glycerol Solutions Containing Free GLDH and NAD^+

Next, we examined the effect of pH (8–10) on the amperometric responses of different glycerol concentrations over a range of 0.04 to 0.20 mM (Figure 6). All the solutions were stirred at a fixed speed and contained 10 units of GLDH and 5 mM NAD^+, and the solution temperature was controlled at 30 °C. Figure 5 shows the current response–concentration profiles obtained with these solutions, which indicates that pH 10 produces a slightly higher current than pH 9, with pH 8 producing the lowest response. It was shown that the response–concentration profiles obtained with these solutions follow Michaelis Menten's behaviour. We also analysed the data using Hanes plots, which confirmed Michaelis Menten's behaviour (Figure S1). Enzymes generally have bell-shaped pH profiles with an optimal pH in the middle, showing increased activity as pH rises initially, followed by decreased activity with further pH increase. The results obtained by Ali & El-Ghonemy [21] indicated an optimum pH for NAD^+ at pH 8–9. From these results, we concluded that pH 9 would be appropriate as the supporting electrolyte for further studies with our amperometric biosensor.

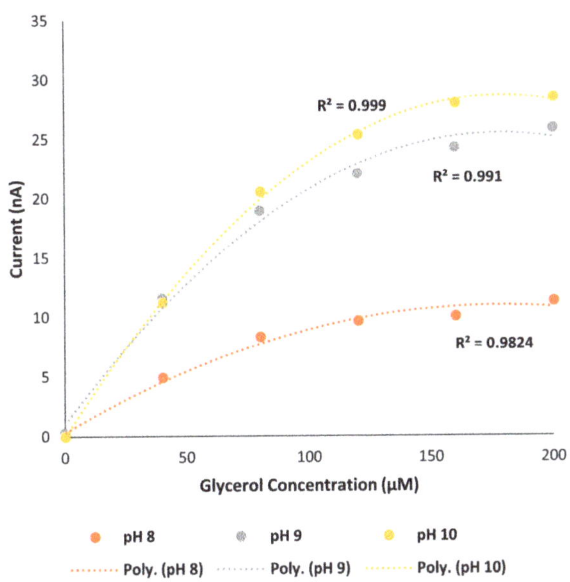

Figure 6. Plot of amperometric responses versus glycerol concentration, obtained with an MB-SPCE in solutions containing glycerol, GLDH (10 U), and NAD^+ (5 mM) at an applied potential of 0.0 V vs. Ag/AgCl.

3.3.3. Calibration Study

A calibration study was performed with the glycerol biosensor in conjunction with chronoamperometry (Figure 7). For this purpose, a 100 µL aliquot of the standard glycerol solutions prepared in 0.1 M PBS pH 9.0 was deposited on top of the working/reference electrodes, and the concentration range of glycerol was 1 to 7.5 mM. Chronoamperometry was performed using an applied potential of 0.0 V vs. Ag/AgCl, preceded by an incubation time of 3 min. Figure 7A,B show the chronoamperomgrams obtained for the standard solutions.

The current was measured at 30 s, 100 s, 200 s, 300 s, and 340 s for each concentration and plotted against the corresponding concentrations. The limits of detection (LOD) were calculated based on the standard error of the regression line (STE_{YX}) of the current response and the slope of the calibration curve (S) according to the formula: LOD = 3.3 × (STE_{YX}/S).

Table 1 shows the linear regression equations, R^2 values, and the LOD for every current sampling time. Based on this data, we considered that a measurement time of 100 s would provide a reliable response for the analytical application described in the next section.

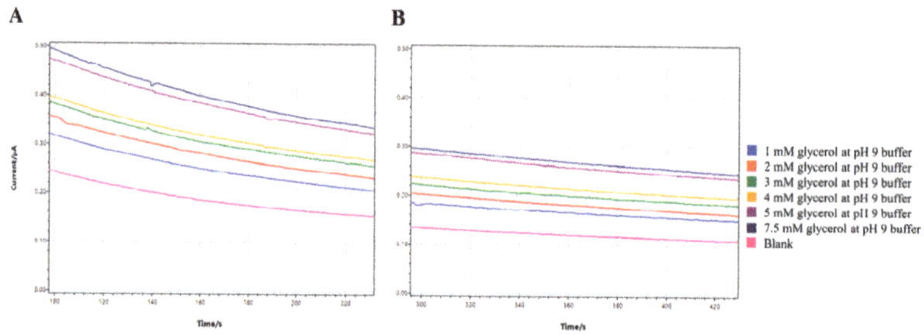

Figure 7. Chronoamperometric responses obtained with the glycerol biosensor for concentrations between 1 and 7.5 mM at (**A**) 100–200 s and (**B**) 300–400 s.

Table 1. The calibration equation, R^2 values, and LOD were obtained using chronoamperometry with the glycerol biosensor at different current sampling times.

Time (s)	Equation	R^2	LOD (mM)
30	y = 0.0317x + 0.4676	0.8916	1.76
100	y = 0.0263x + 0.2643	0.9597	0.87
200	y = 0.0200x + 0.1879	0.9321	1.06
300	y = 0.0181x + 0.143	0.9673	1.25
340	y = 0.0156x + 0.11392	0.9557	1.73

3.4. Analytical Application

The influence of different levels of glycerol on the overall quality of wine continues to be a contentious subject. Previous reports have relied on expensive and time-consuming methods such as phosphorus Nuclear Magnetic Resonance (p-NMR), Gas Chromatography-Mass Spectrometry (GC-MS), and HPLC to detect and quantify glycerol in red wine [22,23].

Table 2 presents the glycerol content recovered from commercial red wine compared to a standard spectrophotometric assay kit for glycerol determination. The mean glycerol concentration detected in red wine (Merlot) using the MB-SPCE glycerol biosensors (11.88 gL^{-1}) was hypothesised not to be different than the glycerol concentration detected using the standard spectrophotometric technique (11.91 gL^{-1}). This hypothesis was confirmed since there was no significant difference between the mean values obtained by the two different methods ($p > 0.05$). The above results demonstrate that our developed MB-SPCE glycerol biosensors can be used as a rapid and efficient analytical method for the determination of glycerol in wines, using only a 100 µL aliquot, which makes this method suitable also for the large-scale control of alcoholic beverages.

It should be mentioned that there have been other reports describing the successful construction of electrochemical biosensors for the measurement of glycerol (Table 3). The fabrication methods vary, involving different combinations of enzymes, transducers, and immobilisation procedures. While these devices have been shown to be reliable in a laboratory setting, they do not have the attributes required for commercialisation, which, amongst other requirements, are the capability of high-speed mass production at low cost to be attractive to sensor manufacturers and end users. Although biosensors constructed with pencil graphite [24], gold [25], carbon paste-packed tubes [26], and graphite rod substrates [27] have demonstrated innovative approaches, the materials and designs do not lend themselves to an economical fabrication approach. In the approach by Mahadevan & Fernando [25] involving grafting FeS to the cofactor NAD^+ with the enzyme GLDH, the sulphur moiety facilitates the attachment of the complex to the surface of a gold electrode.

However, the complexity of the manufacturing procedure would preclude this approach from being used for commercial exploitation.

Table 2. Recovery of glycerol content (in gL^{-1}) from red wine using a glycerol biosensor and Spectrophotometry.

Sample	Recovery (gL^{-1})	
	Biosensor	Spectrophotometry
1	12.58	11.70
2	12.46	11.94
3	11.74	11.94
4	10.06	12.01
5	12.58	11.94
Mean	11.88 [a]	11.91 [a]
SD	0.964	0.097
CV (%)	8.08	0.81

[a] The lowercase letters indicate no significant differences between methods.

Table 3. Brief fabrication details, operating technique, and performance of glycerol biosensors.

Biosensor Fabrication	Technique	LOD	Linear Range	Potential for Commercialisation	References
GLDH/NAD^+ on a meldolas blue-screen-printed carbon electrode	Chronoamperometry Applied potential 0.0 V vs. AgCl/AgCl	0.87 mM	1.0 to 7.5 mM	Yes	Our approach Appl. Sci. 2024
GLDH/NAD^+/Fe-S on a gold electrode	Amperometry in a stirred solution Applied potential +0.6 V vs. Ag/AgCl	0.43 µM	1 µM to 0.1 mM	No	[25]
Alcohol dehydrogenase from Pseudomonas putida immobilised on graphite rod electrode modified with carbon nanotubes and a redox mediator tetrathiafulvalene.	Amperometry in a stirred solution Applied potential +0.05 V vs. SCE	18 µM	0.05 to 10 mM	No	[27]
GLDH/NAD^+/poly(o-phenylenediamine/carbon paste tube electrode	Amperometry in a stirred solution Applied potential 0.0 V vs. Ag/AgCl	0.43 µM	1 µM to 100 mM	No	[26]
Glycerol kinase/glycerol-3-phosphate nanoparticles (NPs) onto graphene oxide nanoparticles (GrONPs) modified pencil graphite (PG) electrode.	Amperometry Applied potential +0.6 V vs. Ag/AgCl	0.002 µM	1µM to 60 mM	No	[24]

4. Conclusions

This paper has shown proof-of-principle studies to demonstrate the feasibility of fabricating a functioning amperometric glycerol biosensor based on an MB-SPCE with immobilised GLDH and NAD^+. This device is convenient to use due to its planar geometry, which permits a microlitre volume of liquid sample or standard solution to be deposited on the biosensor's surface and interrogated using a chronoamperometric procedure. Such chronoamperometric methods can nowadays be performed with small, commercial handheld potentiostats, which allow samples to be analysed outside of the laboratory.

Our approach utilises the most popular industrial method for fabricating biosensors, screen printing with low-cost carbon materials. It should be noted that the one-shot glucose biosensors for diabetes monitoring are fabricated using screen-printing technology, and nowadays, these biosensors have the greatest market share. A simple, single-step enzyme deposition method readily converts the resulting planar transducers into biosensors. In conclusion, our developed MB-SPCE glycerol biosensors can be used as a rapid and efficient analytical method for determining glycerol in wines. This makes the current method suitable for commercialisation and applicable for the large-scale control of alcoholic beverages.

In the future, we will investigate approaches to extend the linear range beyond the limits (1–7.5 mM) described in this paper. This should be readily feasible by adjusting the enzyme and co-factor loadings as well as the biosensor operation. This can help form a platform for developing a novel rapid technology for point-of-test evaluation of glycerol and to assess the practical applicability of the MB-SPCE glycerol biosensors in real-world wine production and quality.

Supplementary Materials: The following supporting information can be downloaded at https://www.mdpi.com/article/10.3390/app14146118/s1, Figure S1: Hanes plots obtained for glycerol solutions under alkaline conditions, with GLDH and NAD^+ in free solution.

Author Contributions: Conceptualisation, S.I.E., A.C., O.D. and J.P.H.; formal analysis, S.I.E., A.C., O.D. and J.P.H.; investigation, S.I.E., A.C., O.D. and J.P.H.; methodology, S.I.E., A.C., O.D. and J.P.H.; project administration, S.I.E., A.C., O.D. and J.P.H.; supervision, S.I.E., A.C., O.D. and J.P.H.; validation, S.I.E., A.C., O.D. and J.P.H.; visualisation, S.I.E., A.C., O.D. and J.P.H.; writing—original draft preparation, S.I.E., A.C., O.D. and J.P.H.; writing—review and editing, S.I.E., A.C., O.D. and J.P.H. All authors have read and agreed to the published version of the manuscript.

Funding: This research received no external funding.

Institutional Review Board Statement: Not applicable.

Informed Consent Statement: Not applicable.

Data Availability Statement: The raw data supporting the conclusions of this article will be made available by the authors on request.

Acknowledgments: The authors would like to thank the University of the West of England (UWE) Bristol for support and Gwent Electronic Materials Ltd., part of the Sun Chemical Corporation, for supplying the screen-printed carbon electrodes.

Conflicts of Interest: The authors declare no conflicts of interest.

References

1. Pagliaro, M. Chapter 1. Glycerol: Properties and Production. In *The Future of Glycerol: New Uses of a Versatile Raw Material;* Mario Pagliaro, M., Rossi, M., Eds.; Royal Society of Chemistry, Green Chemistry Book Series: Cambridge, UK, 2008; pp. 1–17.
2. Kupina, S.; Roman, M. Determination of Total Carbohydrates in Wine and Wine-like Beverages by HPLC with a Refractive Index Detector: First Action 2013.12. *J. AOAC Int.* **2014**, *97*, 498–505. [CrossRef]
3. Li, G.; Han, X.; Liu, F.; Kun-Farkas, G.; Kiss, Z. Simple HPLC Method for Determining the Glycerol Content of Beer. *J. Am. Soc. Brew. Chem.* **2015**, *73*, 314–317. [CrossRef]
4. Lopez, E.F.; Gomez, E.F. Simultaneous Determination of the Major Organic Acids, Sugars, Glycerol, and Ethanol by HPLC in Grape Musts and White Wines. *J. Chromatogr. Sci.* **1996**, *34*, 254–257. [CrossRef]
5. Calull, M.; López, E.; Marcé, R.M.; Olucha, J.C.; Borrull, F. Optimization of an Ion-Exchange High-Performance Liquid Chromatographic Method for the Determination of Carboxylic Acids, Sugars, Glycerol and Ethanol in Wines. *J. Chromatogr. A* **1992**, *589*, 151–158. [CrossRef]
6. Moro, E.; Majocchi, R.; Ballabio, C.; Molfino, S.; Restani, P. A Rapid and Simple Method for Simultaneous Determination of Glycerol, Fructose, and Glucose in Wine. *Am. J. Enol. Vitic.* **2007**, *58*, 279–282. [CrossRef]
7. Edelmann, A.; Diewok, J.; Baena, J.R.; Lendl, B. High-Performance Liquid Chromatography with Diamond ATR–FTIR Detection for the Determination of Carbohydrates, Alcohols and Organic Acids in Red Wine. *Anal. Bioanal. Chem.* **2003**, *376*, 92–97. [CrossRef] [PubMed]
8. Linget, C.; Netter, C.; Heems, D.; Vérette, E. On-Line Dialysis with HPLC for the Automated Preparation and Analysis of Amino Acids, Sugars and Organic Acids in Grape Juice and Wines. *Analusis* **1998**, *26*, 35–39. [CrossRef]
9. Hughes, G.; Westmacott, K.; Honeychurch, K.; Crew, A.; Pemberton, R.; Hart, J. Recent Advances in the Fabrication and Application of Screen-Printed Electrochemical (Bio)Sensors Based on Carbon Materials for Biomedical, Agri-Food and Environmental Analyses. *Biosensors* **2016**, *6*, 50. [CrossRef] [PubMed]

10. Smart, A.; Crew, A.; Doran, O.; Hart, J.P. Studies Towards the Development of a Novel, Screen-Printed Carbon-Based, Biosensor for the Measurement of Polyunsaturated Fatty Acids. *Appl. Sci.* **2020**, *10*, 7779. [CrossRef]
11. Sprules, S.D.; Hart, J.P.; Wring, S.A.; Pittson, R. Development of a Disposable Amperometric Sensor for Reduced Nicotinamide Adenine Dinucleotide Based on a Chemically Modified Screen-Printed Carbon Electrode. *Analyst* **1994**, *119*, 253. [CrossRef]
12. Sprules, S.D.; Hart, J.P.; Wring, S.A.; Pittson, R. A Reagentless, Disposable Biosensor for Lactic Acid Based on a Screen-Printed Carbon Electrode Containing Meldola's Blue and Coated with Lactate Dehydrogenase, NAD+ and Cellulose Acetate. *Anal. Chim. Acta* **1995**, *304*, 17–24. [CrossRef]
13. Sprules, S.D.; Hart, J.P.; Pittson, R.; Wring, S.A. Evaluation of a New Disposable Screen-printed Sensor Strip for the Measurement of NADH and Its Modification to Produce a Lactate Biosensor Employing Microliter Volumes. *Electroanalysis* **1996**, *8*, 539–543. [CrossRef]
14. Sprules, S.D.; Hartley, I.C.; Wedge, R.; Hart, J.P.; Pittson, R. A Disposable Reagentless Screen-Printed Amperometric Biosensor for the Measurement of Alcohol in Beverages. *Anal. Chim. Acta* **1996**, *329*, 215–221. [CrossRef]
15. Hart, J.P.; Abass, A.K.; Cowell, D.C.; Chappell, A. Development of a Disposable Amperometric NH Biosensor Based on a Chemically Modified Screen-Printed Carbon Electrode Coated with Glutamate Dehydrogenase, 2-Oxoglutarate, and NADH. *Electroanalysis* **1999**, *11*, 406–411. [CrossRef]
16. Hart, J.P.; Serban, S.; Jones, L.J.; Biddle, N.; Pittson, R.; Drago, G.A. Selective and Rapid Biosensor Integrated into a Commercial Hand-Held Instrument for the Measurement of Ammonium Ion in Sewage Effluent. *Anal. Lett.* **2006**, *39*, 1657–1667. [CrossRef]
17. Hughes, G.; Pemberton, R.M.; Fielden, P.R.; Hart, J.P. Development of a Disposable Screen Printed Amperometric Biosensor Based on Glutamate Dehydrogenase, for the Determination of Glutamate in Clinical and Food Applications. *Anal. Bioanal. Electrochem* **2014**, *6*, 435–449.
18. Hughes, G.; Pemberton, R.M.; Fielden, P.R.; Hart, J.P. Development of a Novel Reagentless, Screen-Printed Amperometric Biosensor Based on Glutamate Dehydrogenase and NAD+, Integrated with Multi-Walled Carbon Nanotubes for the Determination of Glutamate in Food and Clinical Applications. *Sensors Actuators B Chem.* **2015**, *216*, 614–621. [CrossRef]
19. Westmacott, K.L.; Crew, A.P.; Doran, O.; Hart, J.P. Novel, Rapid, Low-Cost Screen-Printed (Bio)Sensors for the Direct Analysis of Boar Taint Compounds Androstenone and Skatole in Porcine Adipose Tissue: Comparison with a High-Resolution Gas Chromatographic Method. *Biosens. Bioelectron.* **2020**, *150*, 111837. [CrossRef] [PubMed]
20. Lović, J.; Stevanović, S.; Andjelković, B.; Petrović, S.; Vuković, D.; Prlainović, N.; Mijin, D.; Nikolić, N.D.; Ivić, M.A. Electrochemical Glucose Biosensor with the Characterization of Surface Morphology and Content of Glucose Oxidase-Glutaraldehyde-Cysteine Layers on Gold Electrode. *Int. J. Electrochem. Sci.* **2018**, *13*, 12340–12348. [CrossRef]
21. Ali, T.H.; El-Ghonemy, D.H. Purification and Characterization of the Enzymes Involved in Nicotinamide Adenine Dinucleotide Degradation by Penicillium Brevicompactum NRC 829. *3 Biotech* **2016**, *6*, 34. [CrossRef]
22. Hatzakis, E.; Archavlis, E.; Dais, P. Determination of Glycerol in Wines Using 31 P-NMR Spectroscopy. *J. Am. Oil Chem. Soc.* **2007**, *84*, 615–619. [CrossRef]
23. Garcia-Viñola, V.; Ruiz-de-Villa, C.; Gombau, J.; Poblet, M.; Bordons, A.; Reguant, C.; Rozès, N. Simultaneous Analysis of Organic Acids, Glycerol and Phenolic Acids in Wines Using Gas Chromatography-Mass Spectrometry. *Foods* **2024**, *13*, 186. [CrossRef] [PubMed]
24. Narwal, V.; Pundir, C.S. Development of Glycerol Biosensor Based on Co-Immobilization of Enzyme Nanoparticles onto Graphene Oxide Nanoparticles Decorated Pencil Graphite Electrode. *Int. J. Biol. Macromol.* **2019**, *127*, 57–65. [CrossRef] [PubMed]
25. Mahadevan, A.; Fernando, S. An Improved Glycerol Biosensor with an Au-FeS-NAD-Glycerol-Dehydrogenase Anode. *Biosens. Bioelectron.* **2017**, *92*, 417–424. [CrossRef] [PubMed]
26. Álvarez-González, M.I.; Saidman, S.B.; Lobo-Castañón, M.J.; Miranda-Ordieres, A.J.; Tuñón-Blanco, P. Electrocatalytic Detection of NADH and Glycerol by NAD + -Modified Carbon Electrodes. *Anal. Chem.* **2000**, *72*, 520–527. [CrossRef]
27. Ramonas, E.; Ratautas, D.; Dagys, M.; Meškys, R.; Kulys, J. Highly Sensitive Amperometric Biosensor Based on Alcohol Dehydrogenase for Determination of Glycerol in Human Urine. *Talanta* **2019**, *200*, 333–339. [CrossRef]

applied sciences

MDPI

Article

The Effect of Accelerated Storage Temperature Conditions on the Shelf Life of Pasteurized Orange Juice Based on Microbiological, Physicochemical, and Color Attributes

Theofilos Frangopoulos, Antonios Koliouskas and Dimitrios Petridis *

Department of Food Science and Technology, Alexander Campus, International Hellenic University, 57400 Thessaloniki, Greece; thfrangopoulos@gmail.com (T.F.); antonis145159@gmail.com (A.K.)
* Correspondence: petridis@ihu.gr; Tel.: +30-2310013917 or +30-696973336861

Abstract: The accelerated life testing (ASLT) method was used to evaluate the effect of increasing the storage temperature from 10 to 40 °C on the aerobic plate count (APC), the pH, and the colorimetric parameters (L*, a*, b*) of pasteurized orange juice during 40 days of storage. For APC growth, a polynomial model was found to fit better, and at the lower temperatures of 10 and 15 °C, the shelf life was longer, as expected. More specifically, 15 and 10 days were needed, respectively, until the rise in the APC population to 1000 cfu/mL. However, for the temperature range of 30–40 °C, only approximately 3 days were needed to reach 1000 cfu/mL APC. Regarding pH, according to an exponential 3P model, a stable trend was apparent at all temperatures until 30 days of storage, followed by a more abrupt decreasing trend at 25 °C. The lightness (L*), redness (a*), and yellowness (b*) of the juice showed a decreasing trend with the temperature increase, and this trend was more profound at higher temperature levels. The multiple regression analysis between the predictors L*, a*, b*, pH, storage temperature, and the APC response showed an increase in APC growth when the colorimetric parameters decreased and the temperature increased.

Keywords: accelerated life testing (ASLT); orange juice; shelf life; storage temperature; microbiological evaluation; colorimetric parameters

1. Introduction

Orange juice is undoubtedly one of the most commonly consumed products daily worldwide, and thus, it is one of the most commonly exported and imported food products [1]. The pioneer in orange juice production is Brazil, followed by the US, Mexico, and Europe [2]. Moreover, the consumption of orange juice is accompanied by numerous health benefits, including the fact that it is a rich source of vitamin C (ascorbic acid) and a source of many bioactive compounds like carotenoids, flavonoids, folate, and potassium [3]. Despite its many nutritional benefits, orange juice is also a very popular beverage because of its distinguished aroma, flavor, and freshness [4].

However, there is a great necessity to increase the lifetime of the products produced—as they may have to travel a long distance and be stored for a long time to reach their destination—without jeopardizing their safety or their physicochemical and sensory qualities. The most common process through which the safety of the distributed orange juices is guaranteed is thermal pasteurization, which mainly focuses on the inactivation of spoilage and pathogenic microbes and thermally resistant endogenous enzymes, like peroxidase, polyphenol oxidase, and polymethylesterase, which can lead to enzymatic browning [5,6]. However, the thermal processing of orange juice is also correlated with a series of negative sensory attributes, such as color and ascorbic acid retention [6]. Normally, pasteurized orange juice is considered a microbiologically stable product, but several sporoforming bacteria of the genus *Alicyclobacillus* and *Bacillus*, and more specifically *Alicyclobacillus aci-*

doterrestris, *Bacillus coagulans*, and *Bacillus cereus*, can survive pasteurization and germinate in a beneficial environment, thus leading to the deterioration of the product [7].

Nevertheless, apart from the safety point of view, it is also very important to maintain the physicochemical and sensory attributes of the product because they determine consumer acceptability. Among these, the color of orange juice is a significant attribute that is correlated with the freshness and acceptability of the product; so, the handling of the browning reactions that occur during storage before the expiration date is of high importance [8]. The main browning reactions that happen during the storage of pasteurized orange juice are enzymatic and non-enzymatic browning reactions, including the degradation of ascorbic acid, acid-catalyzed degradation of sugars, and Maillard reactions [9]. The kinetics of non-enzymatic browning reactions, which are more difficult to handle, depend on both the physicochemical attributes of products and storage conditions, such as temperature and pH, and more specifically, a higher temperature and higher pH, which intensify these reactions [10].

Under normal conditions, the mean temperature of production and distribution of pasteurized orange juice does not exceed 7 °C, but temperature fluctuations during storage are the main problem that affects the product's acceptability [11]. So, it is very challenging when estimating the shelf life of products to be able to incorporate these unwanted deviations of conditions. Moreover, the shelf life and expiration date estimation are based on microbiological safety and not sensorial and physicochemical characteristics or "sensory shelf life", which poses the risk that the product is considered safe for consumption but its sensory characteristics have been compromised, which has a negative impact on consumer acceptability [12].

A significant innovation in the field of the prediction of food shelf life is accelerated shelf life testing (ASLT), which provides the possibility of predicting the shelf life of products based on both microbial and sensory characteristics during different stressing conditions, like temperature, without extensive amounts of testing and time using different statistical models [13]. In the field of perishable types of food products, like fruit juices, many studies have been conducted using temperature, UV radiation, and others as acceleration factors during different storage periods. These studies mainly focused on the effect of stress conditions on microbial counts, color deterioration, and unwanted oxidation reactions, thus mainly using Arrhenius or Weibull equations [4,5,14–17].

Meanwhile, the present study aims to fill the gap at the research and industrial levels regarding the correlation between the microbial, physicochemical, and color attributes of pasteurized orange juice stored at different temperatures for an extensive amount of time. Moreover, it is the first instance in the literature in which the data used are not from laboratory-produced samples but industrial samples from the actual production line.

2. Materials and Methods

2.1. Materials

Commercial pasteurized orange juice was used in the present study, which was manufactured at a local juice production factory in 500 mL PET bottles. Directly after production, the pasteurized orange juice bottles were stored at different temperatures in the company's industrial refrigerators.

2.2. Methods

2.2.1. Physicochemical Parameters

The pH of the pasteurized orange juice was measured using a bench pH meter (Mettler Toledo FP20, Columbus, OH, USA). The pH of the samples was measured after shaking the flask for 15 s, directly from the flask and in triplicate (identical replicates). For temperature equilibration, before measuring the pH, the flasks were left for 1 h at room temperature. Each day, the calibration of the caliper was carried out with pH 4 and pH 7 buffer.

2.2.2. Color Parameters

The chroma parameters (L*, a*, and b*) were evaluated using the chromameter Lovibond LM200 (Lovibond, Amesbury, UK). The parameters L*, a*, and b* were measured after shaking the bottle for 15 s by taking a 0.5 mL sample directly from the bottle and in triplicate (identical replicates).

2.2.3. Microbiological Parameters

The automatic analyzer BioMérieux Tempo (bioMérieux, Marcy-l'Étoile, France) was used for the evaluation of the total viable count of the orange juice samples. More specifically, the method is based on ISO 16140:2003 [18] protocol for microbiology of food and animal feeding stuffs. The method manipulates the automated MPN (most probable number). The process of dilutions is performed inside the Tempo card. The Tempo card carries 3 rows of 16 wells each (48 wells in total) corresponding to 3 dilutions with 16 tubes each. The upper MPN threshold of the automatic analyzer at a sample without dilution is 4900. The microbial growth was measured daily from three 500 mL bottles (uniform replicates) until the aerobic plate count (APC) value approached the maximum value of 4900 units, which marked the end of the experimental run.

2.2.4. Statistical Analysis

The temperature was used as an accelerating degradation factor choosing 6 levels (10, 15, 20, 25, 30, and 40 °C) to monitor the APC growth and 4 levels (4, 15, 20, and 25) to plot the three colorimetric parameters and pH with time (days). Several regression equation models were used to find the best fit of the experimental data using AICc and BIC as information criteria and the determinant coefficient R^2 to determine the model reliability. All statistical analyses were performed via JMP 17 (Statistical Discovery LLC, Cary, NC, USA) software.

All models were based on empirical choice since no recent information exists in literature relevant to some fundamental work. To determine microbial growth, five models were used, all of which involved log transformation of the APC response. An exponential equation with 3 parameters was used for pH modeling, a logistic equation with 3 parameters was used for Lightness L*, and two linear equations were used for Redness a* and Yellowess b*.

3. Results and Discussion

3.1. Aerobic Plate Count (APC) of Pasteurized Orange Juice

The parsimonious polynomial model below with only four parameters (4P) best describes the APC growth rate (*N*) over the storage time (*t*) at various temperature levels (*T*) along with the statistical parameters of the model (Table 1):

$$N = N_0 + \exp(b_1 + b_2 * Ln(T)) * t \qquad (1)$$

Table 1. Statistical information for APC model.

Parameter	Estimate	Std Error	−95%	95%
b0	−0.238380	0.40665308	−1.035405	0.558645
b1	−3.550622	0.22067326	−3.983134	−3.118111
b2	1.182725	0.06742677	1.050571	1.314880
Sigma	1.130722	0.12053435	0.894479	1.366965

The strong linear shape of standardized and Cox–Snell residuals indicated a very reliable model (AICc = 607.55).

Figure 1 depicts the population of APC (cfu/mL) versus storage time at each storage temperature. As can be seen, at lower temperatures (10 °C and 15 °C), the shelf life is longer than expected, and more specifically, 15 and 10 days are needed, respectively, until the rise of the population to the industrial internal limit of 1000 cfu/mL APC. The APC values converge at 20–25 °C and at 30–40 °C, indicating that at the temperature range of 20–25 °C, 6.5 days are needed to reach 1000 cfu/mL, whereas, at the same time, for the temperature range of 30–40 °C, only 3.5 days are needed to reach 1000 cfu/mL. The steep increase in the growth rate following the rise in the storage temperature was also observed in work by Caranza et al. (2021) [15], where the Weibull equation was used to model the shelf life of orange juice at different temperature levels. At 5 °C and 22 °C, the inactivation rate, b, of the equation was 0.01 and 3.88, respectively, while in the present study, the same values were 2.18 and 5.67, respectively. Although a direct comparison between the two studies is not feasible due to the different equations (Weibull and polynomial), a rough estimation could be informative. The premature and rapid increase in APC at the higher temperature levels (20–40 °C) is explained by the diminishing of the lag phase at storage temperatures above 15 °C, meaning that this decrease in the lag phase, where the microbial growth rate is zero, consequently leads to a reduction in shelf life due to faster spoilage [17]. More specifically, in the study performed by Raccach and Mellattdoust [17], it was observed that the storage of orange juice at different temperature levels (0, 4, 10, and 15 °C) reduced the APC lag phase period from 9.5 days to 2.8 and 0 days for 10 and 15 °C, respectively. In the present study, this decrease in the APC lag phase, along with the increase in the storage temperature, was also apparent. As a matter of fact, in Figure 1, the lag phase at temperatures of 10, 15, 20, and 25 °C is 8, 4, 2, and 1 day, respectively [19].

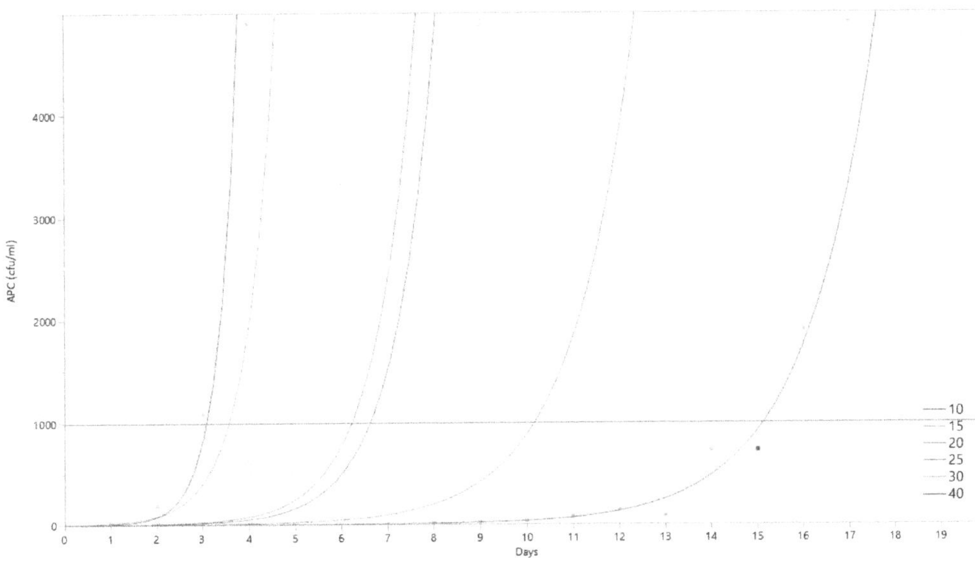

Figure 1. Population changes of APC as a function of storage time in different temperature levels.

The advantage of the polynomial model is the perception of shelf life as a continuous and not categorical variable, enabling the prediction of the shelf life at different temperature values than the experimented ones. This is very important when considering that the usual storage temperatures of pasteurized orange juice do not normally exceed 7 °C. Thus, using the degradation profiler (Figure 2), which depicts the APC growth at different time points (days) and temperature levels (°C), the required amount of time for achieving 1000 cfu/mL (internal limit of expiration) at different temperature levels is smoothly obtained. Briefly, if the temperature level of Figure 2 is adjusted at different levels, the predictive table (Table 2) can be created, presenting the shelf life of pasteurized orange juice. In this table, it can be observed that orange juice can be safely consumed for 48, 37, and 30 days if preserved at 4, 5, and 6 °C. However, when the temperature reaches 10 °C, safe consumption is reduced to 16 days, meaning that the 4-degree rise in temperature leads to a 14-day loss of shelf life. In this context, in the study by Supraditareporn and Pinthong [20] a 6-day loss of shelf life of fresh orange juice was observed when the storage temperature was shifted from 4 to 25 °C. Increasing the storage temperature levels leads to a further reduction in the shelf life, as can also be seen in Figure 1, reaching approximately 3 days at 40 °C. In a similar study, where thermally treated sweet lime juice was subjected to accelerated storage conditions at 4, 15, and 25 °C, the time needed to reach 1000 cfu/mL or 3 logs cfu/mL was ~42, 25, and 12 days, respectively [16]. In the present study, the shelf life under these conditions was 48, 10, and 6 days.

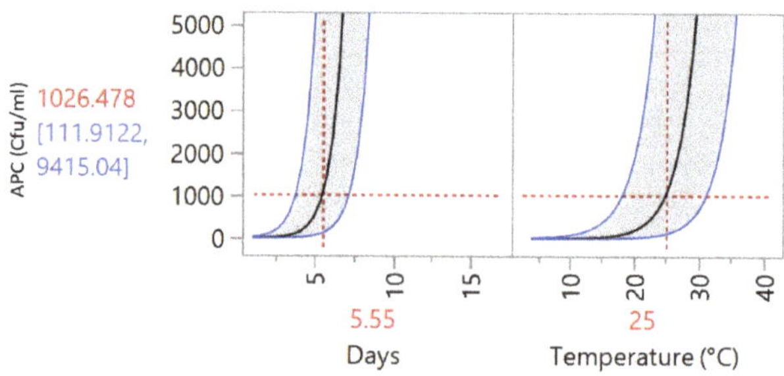

Figure 2. Degradation profiler of APC versus period (days) and temperature level using a polynomial model.

Table 2. Shelf life of pasteurized orange juice at different temperature levels.

Temperature (°C)	Days	APC (Cfu/mL)	−95%	95%
4	48.31	1000.4	109.1	9175.5
5	37.11	1001.5	109.2	9186.3
6	30.03	-	-	-
10	16.35	1002.6	109.3	9195.8
15	10.12	1001.4	109.2	9185.3
20	7.2	1000.1	109	9173.2
25	5.53	1000.3	109.1	9174.8
30	4.46	1001.4	109.2	9184.6
40	3.17	1000.7	109.1	9178.7

3.2. pH of Pasteurized Orange Juice

The following exponential 3P model describes the best-fit change in pH with the storage time via temperature as an accelerating factor (AICc = 418.2, R^2 = 0.554):

$$Y = a + b * e^{c*Days} \quad (2)$$

where a is the asymptote, b is the scale, and c is the growth rate.

Table 3 presents comparisons of the inflection points between different aging temperatures of the juice pH. This table highlights the decreasing trend of the pH along with the temperature increase based on the increasing growth rates, although this trend is more evident and statistically significant at higher temperature levels (20–25 °C) based on p-values. Furtherly, the progression of the pH is stable during the storage time at levels ranging between 3.9 and 4 until the 30th day, regardless of the storage temperature (Figure 3). This trend was also observed in the work performed by Wahia et al. [21], where the pH of orange juice remained nearly stable during 24 days of storage at 4 and 25 °C, with values ranging from 4.45 to 4.23 and 4.31 to 4.29, respectively. Also, in the work performed by Cortes et al. (2008) [4], the pH of pasteurized orange juice remained stable for 6 weeks at 2 and 10 °C, with values ranging from 3.32 to 3.49 and 3.32 to 3.48, respectively. In our study, from the 30th to the 40th day of storage, the pH values declined progressively, with greater intensity for the samples with a higher storage temperature. More specifically, regarding samples preserved at 25 °C, a steep decrease in the pH from 3.8 to 3.4 was observed from the 30th to the 40th day (Figure 3), which is also illustrated by the value of the specific growth rate, i.e., 0.31, compared to the growth rates of 4 °C and 15 °C, which were 0.18 and−0.09, respectively (Table 3). This trend could be attributed to the rapid metabolization of the orange juice simple carbohydrates by the large numbers of aerobic plate counts at prolonged storage periods and higher temperatures, leading to the accumulation of organic acids as by-products of the reaction, subsequently lowering the pH values [21,22]. Under normal conditions, in a pasteurized product, the bacteria that ferment sugars into organic acids are heat-sensitive and are destroyed, but the spore-forming strains of the *Alicyclobacillus* genera and, more specifically, *Alicyclobacillus acidoterrestris*, survive pasteurization and, after germination in ideal conditions, they cause acidification of the juice [23].

Table 3. Statistical information on pH during different aging conditions.

Temperature (°C)	Asymptote	Scale	Growth Rate	*p*-Value
4	3.959574	−0.017114	−0.085518	0.8688
15	3.9830888	−0.000074	0.1864306	0.1460
20	3.9434177	-4.768×10^{-7}	0.322794	0.048
25	3.949530	-2.297×10^{-6}	0.3089789	<0.0001

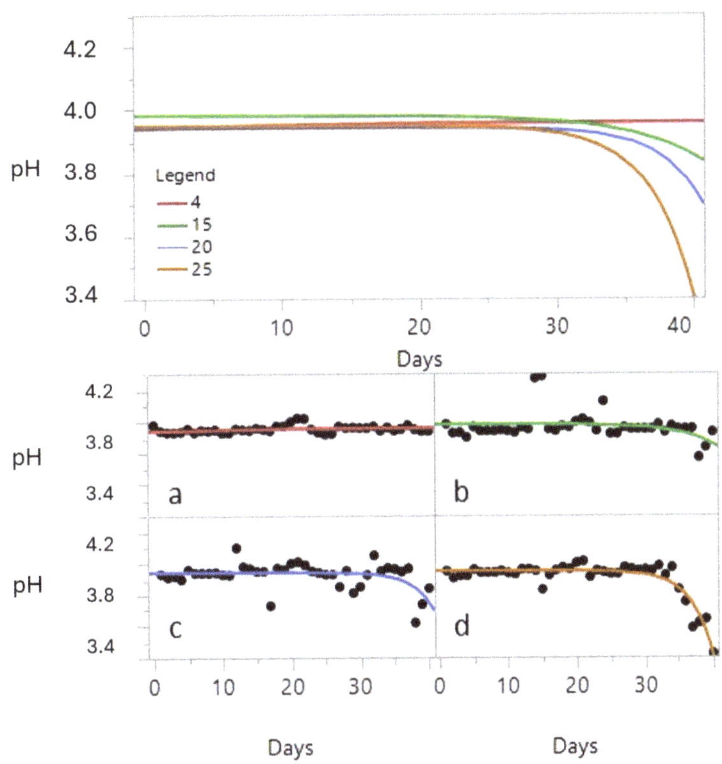

Figure 3. Curve fitting plot of experimental data in the exponential 3P model of pH vs. storage days. (**a**) 4 °C; (**b**) 15 °C; (**c**) 20 °C; (**d**) 25 °C.

3.3. *Color Characteristics of Pasteurized Orange Juice*

Regarding the color characteristics of pasteurized orange juice, L*, a*, and b* were regularly monitored during the storage time at increasing temperature levels to model the degradation process.

3.3.1. Lightness (L*)

The logistic 3P model proved to fit data slightly better than the Gompertz 3P model, as Table 4 indicates (lower AICc and BIC values and higher AICc weight and R^2 values):

$$Y = \frac{c}{1 + e^{-a*(day-b)}} \tag{3}$$

Table 4. Statistical information of the two candidate models used for describing the L* value.

Model	AICc	AICc Weight	BIC	SSE	MSE	RMSE	R-Square
Logistic 3P	465.22	0.556	502.81	142.06	0.953	0.9764	0.7498
Gompertz 3P	465.67	0.444	503.26	142.46	0.956	0.9773	0.7491

In Table 5, a general decreasing trend of L* is observed, based on the negative values of the growth rate in all temperature levels, almost all significant, in a decreasing order per level. Although the reduction in the L* value below 48 initiates at ~30 days of storage at 4 °C, at the higher temperature levels (15, 20, and 25 °C), the reduction initiates at ~25, ~20, and <10 days, respectively (Figure 4). These results follow the trend in the literature, and, more specifically, the study performed by Wahia et al. [21] and Cortes et al. [4], where a decrease in the L* values of juice during storage is apparent, particularly at higher storage temperatures. Of course, the L* value indicates the lightness of orange juice, and, expectedly, during the storage duration, oxidative and enzymatic browning reactions lead to a change in the juice color from brighter to darker [24,25]. However, since most of the browning reactions are intensified at higher temperatures, according to the Arrhenius laws, it is reasonable for a premature and more intense reduction in the L* values to occur as the temperature rises [26]. Another explanation for the browning intensity at higher storage temperatures is the rapid degradation of ascorbic acid at higher temperature conditions. The ascorbic acid acts protectively on the juice color because of its high antioxidant activity; thus, it inhibits the oxidative and non-oxidative reactions that cause browning [27].

Table 5. Statistical information on L* during different aging conditions.

Temperature (°C)	Growth Rate	Inflection Point	Asymptote	*p*-Value
4	−0.241786	51.705908	47.720108	0.0569
15	−0.141779	58.235922	47.872129	0.0134
20	−0.135923	55.134761	47.808139	<0.0001
25	−0.068845	63.301295	48.350857	<0.0001

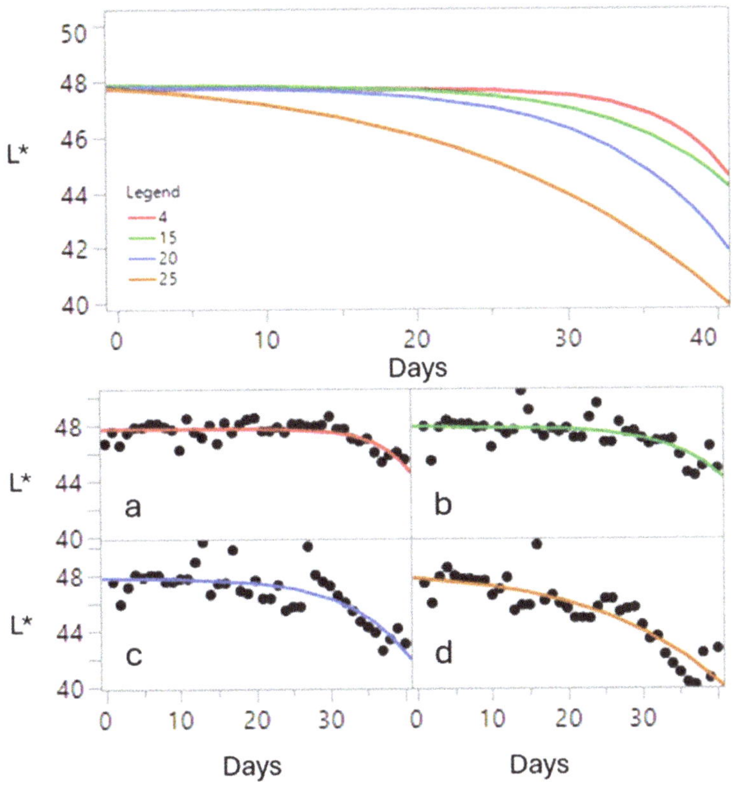

Figure 4. Curve fitting plot of experimental data in the logistic 3P model of L* value vs. storage days. (**a**) 4 °C; (**b**) 15 °C; (**c**) 20 °C; (**d**) 25 °C.

3.3.2. Redness (a*)

Regarding the redness of the juice, a linear model presented the best fit (AICc = 96.97, $R^2 = 0.64$):

$$Y = a - b * day \tag{4}$$

The redness of the pasteurized orange juice declines during storage, as manifested in Figure 4 and statistically described in Table 6. In other words, the rate of the decrease is greatly affected by the storage temperature level. Although at 4 and 15 °C, the reduction was relatively mild, with a change from 6.7 to 6.5, at the higher temperature levels of 20 and 25 °C, the reduction was steeper, and dropped from approximately 7 to 5 at the end of the storage time (Figure 5). This observation was also noticed in the study performed by Choi et al. [28], where the a* parameter of blood orange juice decreased from 17.5 to 15.4 during storage for 7 weeks, and this was associated with both the degradation of anthocyanins, which are the main color agents of orange juice, and ascorbic acid. Anthocyanins, which are polyphenols, present high antioxidant activity and have numerous beneficial effects on human health. Thus, maintaining high levels of anthocyanins after the production of orange juice improves the nutritional value of the product [29]. This fact enhances the relationship between the optical and the nutritional value of orange juice, enabling the establishment of new indirect nutritional indicators.

Table 6. Statistical information on a* during different aging conditions.

Temperature (°C)	Estimate	Std Error	*p*-Value
4	−0.006992	0.0041714	0.0937
15	−0.008871	0.0043288	0.0404
20	−0.032236	0.0043288	<0.0001
25	−0.047405	0.0043288	<0.0001

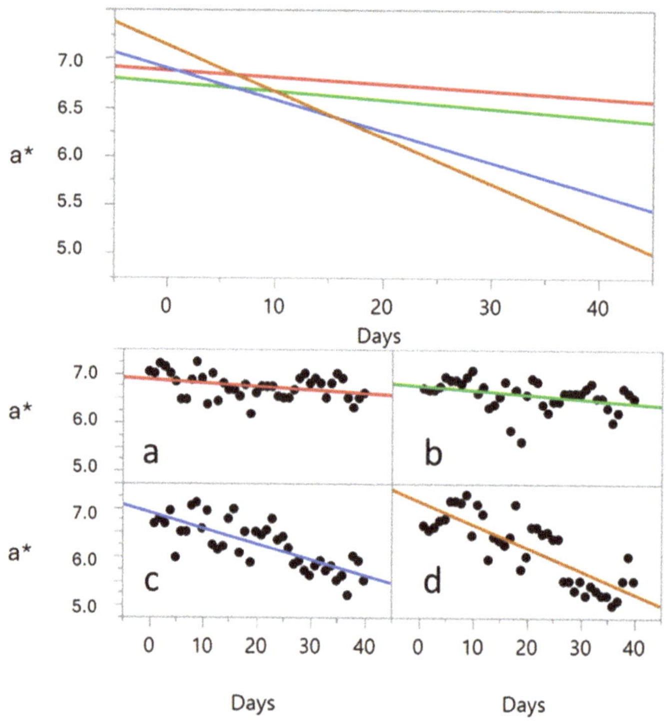

Figure 5. Curve fitting plot of experimental data in the linear model of a* value vs. storage days. (**a**) 4 °C; (**b**) 15 °C; (**c**) 20 °C; (**d**) 25 °C.

3.3.3. Yellowness (b*)

Regarding the yellowness of the juice, a linear model presented the best fit (AICc = 450.48, R^2 = 0.79). The linear model is expressed via the following equation:

$$Y = a - b * day \quad (5)$$

Yellowness is a very important color factor of orange juice because it is correlated with the freshness of the product and thus is linked with higher consumer acceptability [30]. During pasteurization, due to the generation of melanoidins at high temperatures, yellowness is prone to decrease, a process that is more intensive as the storage time passes [31]. Thus, it is very important to maintain the yellowness of pasteurized orange juice at relatively high levels. In the present study, the yellowness slowly decreases at lower temperature levels (4, 15, and 20 °C) at values ranging from ~42 to ~40 at the end of the storage time (Figure 6). At 25 °C, a steep decrease in the yellowness is observed during the storage time, with values ranging from ~42 to ~34 at the end of the storage time (Table 7). These results are in an analog accordance with the literature and, more specifically, with the work performed by

Wahia et al. [21], where, during the storage of pasteurized orange juice for 24 days at 4 and 25 °C, the b* values ranged from 11.02 to 11.15 and 12.91 to 9.34, respectively.

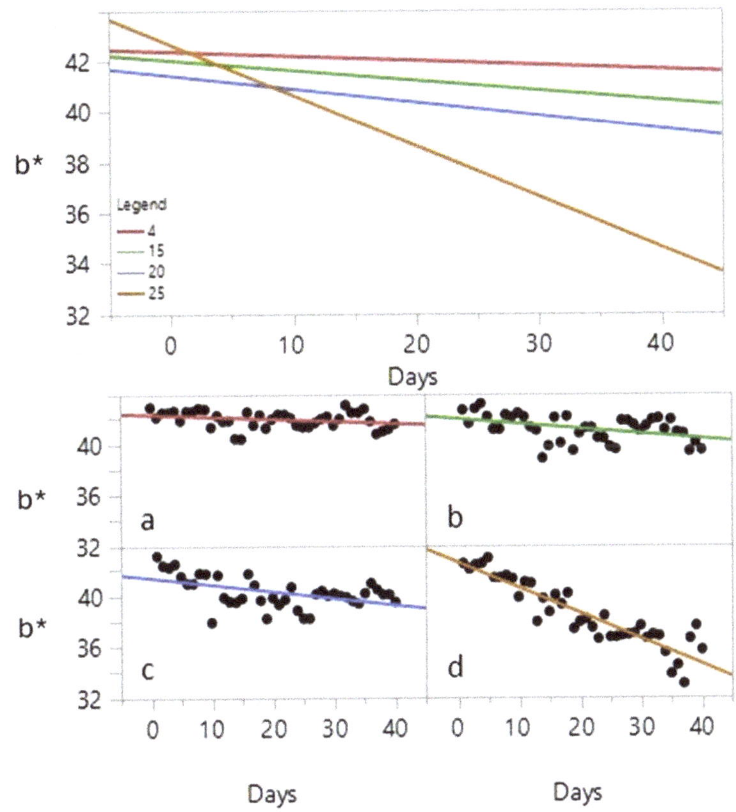

Figure 6. Curve fitting plot of experimental data in the linear model of b* value vs. storage days. (**a**) 4 °C; (**b**) 15 °C; (**c**) 20 °C; (**d**) 25 °C.

Table 7. Statistical information on b*during different aging conditions.

Temperature (°C)	Estimate	Std Error	*p*-Value
4	−0.018118	0.0125045	0.1473
15	−0.039981	0.0129765	0.0021
20	−0.05344	0.0129765	<0.0001
25	−0.201351	0.0129765	<0.0001

3.4. Relationships Between APC and Predictors

The present study not only tried to evaluate the shelf life of pasteurized orange juice as a function of the aerobic plate count (APC) but also from a physicochemical point of view. Thus, at each time point and storage temperature (stress factor), the color (L*, a*, and b*) and pH were measured. However, the microbiological deterioration of orange juice at each temperature level inevitably affects the physicochemical characteristics [32]. It is thus of great interest to examine the relationship between the APC growth and the physicochemical variables at each storage temperature level. The relationship between the APC and the physicochemical variables was examined by using a reformed spreadsheet from the present data: four temperature levels (5, 15, 20, 25) at 10 specific storage days and mainly from the last days of storage were selected, and the predicted APC values were extrapolated via the equation models (as previously explained) for a greater number days of storage for reasons of data conformation.

The results of the multiple regression analysis are shown in the table (Table 8), in which the parameters L*, a*, and temperature significantly affect the aerobic plate count. The prediction profiler of the APC versus all the predictors (Figure 7) shows that the increasing APC growth is related to a decrease in lightness (L*), redness (a*), and yellowness (b*) and an increase in the temperature, as expected. As stated by Wahia et al. [21], changes in the colorimetric L*, a*, and b* during the storage of orange juice are interrelated due to both oxidative and non-oxidative darkening reactions. However, it was observed that the color degradation was also directly related to the increase in the APC values. This is mainly attributed to the activity of sporoforming spoilage bacterial strains, which produce hydrolytic enzymes that cause cloudiness, off-odors, and deterioration of the juices [23].

Table 8. Statistical information on APC response using L*, a*, b*, pH, and temperature as predictors.

Term	Estimate	Std Error	t Ratio	*p*-Value
Intercept	211.08049	35.22313	5.99	<0.0001
L*	−2.478491	0.888081	−2.79	0.0086
a*	−8.901556	2.981804	−2.99	0.0052
Temperature (°C)	0.5427967	0.169693	3.20	0.0030
b*	−0.45272	0.80784	−0.56	0.5789
pH	−3.101464	9.755159	−0.32	0.7525

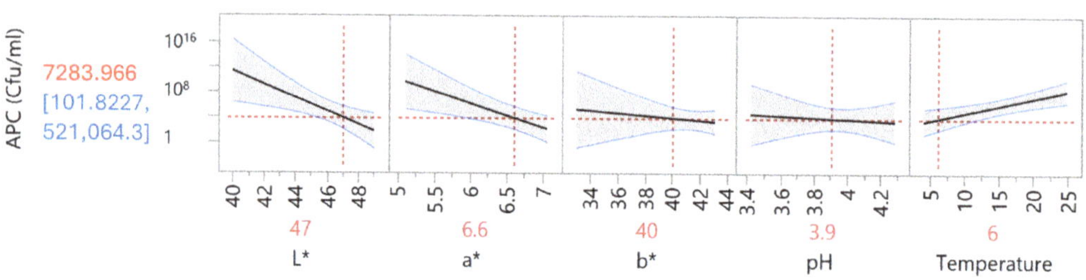

Figure 7. Prediction profiler of APC (cfu/mL) baseline in relation to L*, a*, b*, pH, and temperature level (°C).

The above APC profiler is extremely useful for applications in the industry because, apart from its predictive value, it presents the possibility of using the color characteristics of pasteurized orange juice as indices of the microbiological deterioration at each time point and thus of the shelf life of the products.

4. Conclusions

In this study, pasteurized orange juice was subjected to accelerated storage temperature conditions for 40 days. The tested parameters were the aerobic plate count, the colorimetric parameters (L*, a*, and b*), and pH. The results showed that a polynomial rate model described better the APC growth, which was more intense as the storage temperature level increased. Exponential 3P, logistic 3P, and linear models were found to adequately describe pH, L*, a*, and b* parameters throughout the storage period. A multiple regression analysis using the APC growth as a response against all the aforementioned parameters reveals an APC increase with the downgrading of colorimetric parameters and temperature elevation. Plus, it is very important to highlight the practical aspects of the research and the very important effect of the fluctuations of the preservation temperature in the cold chain on the shelf life of the juice, which enables the improvement in storage conditions. Secondly, this study creates the basis for the creation of microbiological health indicators of juice that can be measured quickly and directly without the need for microbiological analysis and specialized personnel.

Author Contributions: Conceptualization, D.P.; Methodology, D.P. and A.K.; Validation, T.F., D.P. and A.K.; Formal Analysis, D.P., A.K. and T.F.; Investigation, D.P., A.K. and T.F.; Software, D.P. and T.F.; Resources, D.P. and A.K.; Data Curation, D.P.; Writing—Original Draft Preparation, T.F. and D.P.; Writing—Review and Editing, T.F. and D.P.; Visualization, T.F. and D.P.; Supervision, D.P.; Project Administration, D.P.; Funding Acquisition, D.P. All authors have read and agreed to the published version of the manuscript.

Funding: This research received no external funding.

Institutional Review Board Statement: Not applicable.

Informed Consent Statement: Not applicable.

Data Availability Statement: The original contributions presented in the study are included in the article; further inquiries can be directed to the corresponding authors.

Conflicts of Interest: The authors declare no conflicts of interest.

References

1. Charfi, S.; Boujida, N.; Abrini, J.; Senhaji, N. Study of chemical composition and antibacterial activity of Moroccan Thymbra capitata essential oil and its possible use in orange juice conservation. *Mater. Today Proc.* **2019**, *13*, 706–712. [CrossRef]
2. Citrus: World Markets and Trade. 2023. Available online: https://downloads.usda.library.cornell.edu/usda-esmis/files/w66343603/cz30rd902/vh53zg80g/citrus.pdf (accessed on 24 July 2024).
3. Ivanova, N.N.; Khomich, L.M.; Perova, I.B. Orange juice nutritional profile. *Vopr Pitan* **2017**, *86*, 103–113. (In Russian) [PubMed]
4. Cortés, C.; Esteve, M.J.; Frígola, A. Color of orange juice treated by High Intensity Pulsed Electric Fields during refrigerated storage and comparison with pasteurized juice. *Food Control* **2008**, *19*, 151–158. [CrossRef]
5. Wibowo, S.; Grauwet, T.; Kebede, B.T.; Hendrickx, M.; Van Loey, A. Study of chemical changes in pasteurised orange juice during shelf-life: A fingerprinting-kinetics evaluation of the volatile fraction. *Food Res. Int.* **2015**, *75*, 295–304. [CrossRef] [PubMed]
6. Mandha, J.; Shumoy, H.; Matemu, A.O.; Raes, K. Characterization of fruit juices and effect of pasteurization and storage conditions on their microbial, physicochemical, and nutritional quality. *Food Biosci.* **2023**, *51*, 102335. [CrossRef]
7. Colás-Medà, P.; Nicolau-Lapeña, I.; Viñas, I.; Neggazi, I.; Alegre, I. Bacterial Spore Inactivation in Orange Juice and Orange Peel by Ultraviolet-C Light. *Foods* **2021**, *10*, 855. [CrossRef]
8. Fernández-Vázquez, R.; Hewson, L.; Fisk, I.; Vila, D.H.; Mira, F.J.H.; Vicario, I.M.; Hort, J. Colour influences sensory perception and liking of orange juice. *Flavour* **2014**, *3*, 1. [CrossRef]
9. Buvé, C.; Pham, H.T.T.; Hendrickx, M.; Grauwet, T.; Van Loey, A. Reaction pathways and factors influencing nonenzymatic browning in shelf-stable fruit juices during storage. *Compr. Rev. Food Sci. Food Saf.* **2021**, *20*, 5698–5721. [CrossRef]
10. Karseno, E.; Yanto, T.; Setyawati, R.; Haryanti, P. Effect of pH and temperature on browning intensity of coconut sugar and its antioxidant activity. *Food Res.* **2017**, *2*, 32–38. [CrossRef]
11. Khasanov, A.; Matveeva, N. Determination of the shelf life of a functional beverage by accelerated testing. *E3S Web Conf.* **2020**, *164*, 01003. [CrossRef]
12. EFSA; Guidance on date marking and related food information: Part 1 (date marking). *EFSA J.* **2020**, *18*, e06306.
13. Haouet, M.N.; Tommasino, M.; Mercuri, M.L.; Benedetti, F.; Di Bella, S.; Framboas, M.; Pelli, S.; Altissimi, M.S. Experimental accelerated shelf life determination of a ready-to-eat processed food. *Ital. J. Food Saf.* **2018**, *7*, 189–192. [CrossRef] [PubMed]
14. Buvé, C.; Van Bedts, T.; Haenen, A.; Kebede, B.; Braekers, R.; Hendrickx, M.; Van Loey, A.; Grauwet, T. Shelf-life dating of shelf-stable strawberry juice based on survival analysis of consumer acceptance information. *J. Sci. Food Agric.* **2018**, *98*, 3437–3445. [CrossRef] [PubMed]
15. Hernández-Carranza, P.; Peralta-Pérez, A.; Avila-Sosa, R.; Ruiz-López, I.I.; Benitez-Rojas, A.C.; Ochoa-Velasco, C.E. Effect of ultraviolet-C light and mild thermal treatment on the storage life of orange juice. *Czech J. Food Sci.* **2021**, *39*, 106–112. [CrossRef]
16. Shaik, L.; Chakraborty, S. Effect of different storage conditions on the quality attributes of sweet lime juice subjected to pulsed light and thermal pasteurization. *Sustain. Food Technol.* **2023**, *1*, 722–737. [CrossRef]
17. Raccach, M.; Mellatdoust, M. The effect of temperature on microbial growth in orange juice. *J. Food Process. Preserv.* **2007**, *31*, 129–142. [CrossRef]
18. *ISO 16140:2003*; Microbiology of Food and Animal Feeding Stuffs—Protocol for the Validation of Alternative Methods. ISO: Geneva, Switzerland, 2003.
19. Buchanan, R.; Whiting, R.C.; Damert, W.C. When is simple good enough: A comparison of the Gompertz, Baranyi, and three-phase linear models for fitting bacterial growth curves* 1,* 2. *Food Microbiol.* **1997**, *14*, 313–326. [CrossRef]

20. Supraditareporn, W.; Pinthong, R. Physical, Chemical and Microbiological Changes during Storage of Orange Juices cv. Sai Nam Pung and cv. Khieo Waan in Northern Thailand. *Int. J. Agric. Biol.* **2007**, *9*, 726–730.
21. Wahia, H.; Zhou, C.; Mustapha, A.T.; Amanor-Atiemoh, R.; Mo, L.; Fakayode, O.A.; Ma, H. Storage effects on the quality quartet of orange juice submitted to moderate thermosonication: Predictive modeling and odor fingerprinting approach. *Ultrason. Sonochem.* **2020**, *64*, 104982. [CrossRef]
22. Zhu, W.; Ai, Y.; Fang, F.; Liao, H. Application of Thermosonication in Red Pitaya Juice Processing: Impacts on Native Microbiota and Quality Properties during Storage. *Foods* **2021**, *10*, 1041. [CrossRef]
23. Sourri, P.; Tassou, C.C.; Nychas, G.-J.E.; Panagou, E.Z. Fruit Juice Spoilage by *Alicyclobacillus*: Detection and Control Methods—A Comprehensive Review. *Foods* **2022**, *11*, 747. [CrossRef] [PubMed]
24. Polydera, A.; Stoforos, N.; Taoukis, P. Quality degradation kinetics of pasteurised and high pressure processed fresh Navel orange juice: Nutritional parameters and shelf life. *Innov. Food Sci. Emerg. Technol.* **2005**, *6*, 1–9. [CrossRef]
25. Pham, H.T.T.; Kityo, P.; Buvé, C.; Hendrickx, M.E.; Van Loey, A.M. Influence of pH and Composition on Nonenzymatic Browning of Shelf-Stable Orange Juice during Storage. *J. Agric. Food Chem.* **2020**, *68*, 5402–5411. [CrossRef] [PubMed]
26. Fustier, P.; St-Germain, F.; Lamarche, F.; Mondor, M. Non-enzymatic browning and ascorbic acid degradation of orange juice subjected to electroreduction and electro-oxidation treatments. *Innov. Food Sci. Emerg. Technol.* **2011**, *12*, 491–498. [CrossRef]
27. Yin, X.; Chen, K.; Cheng, H.; Chen, X.; Feng, S.; Song, Y.; Liang, L. Chemical Stability of Ascorbic Acid Integrated into Commercial Products: A Review on Bioactivity and Delivery Technology. *Antioxidants* **2022**, *11*, 153. [CrossRef]
28. Choi, M.; Kim, G.; Lee, H. Effects of ascorbic acid retention on juice color and pigment stability in blood orange (*Citrus sinensis*) juice during refrigerated storage. *Food Res. Int.* **2002**, *35*, 753–759. [CrossRef]
29. Marszałek, K.; Woźniak, L.; Kruszewski, B.; Skąpska, S. The Effect of High Pressure Techniques on the Stability of Anthocyanins in Fruit and Vegetables. *Int. J. Mol. Sci.* **2017**, *18*, 277. [CrossRef]
30. Tepper, B.J. Effects of slight color variation on consumer acceptance of orange juice. *J. Sens. Stud.* **1993**, *8*, 145–154. [CrossRef]
31. Aghajanzadeh, S.; Ziaiifar, A.M.; Verkerk, R. Effect of thermal and non-thermal treatments on the color of citrus juice: A review. *Food Rev. Int.* **2021**, *39*, 3555–3577. [CrossRef]
32. Kaddumukasa, P.P.; Imathiu, S.M.; Mathara, J.M.; Nakavuma, J.L. Influence of physicochemical parameters on storage stability: Microbiological quality of fresh unpasteurized fruit juices. *Food Sci. Nutr.* **2017**, *5*, 1098–1105. [CrossRef]

applied sciences

MDPI

Article

Sensory Shelf Life of Pasteurized Orange Juice Stored Under Different Temperature Levels Using Inverse Time Sampling and a Balanced Incomplete Block Design

Theofilos Frangopoulos, Antonios Koliouskas and Dimitrios Petridis *

International Hellenic University, 57400 Sindos, Greece; thfrangopoulos@gmail.com (T.F.); antonis145159@gmail.com (A.K.)
* Correspondence: petridis@ihu.gr; Tel.: +30-2310013917 or +30-696973336861

Abstract: The sensory acceptability rate of pasteurized orange juices stored at different temperature levels was evaluated using a large-scale experiment with untrained consumers and an accelerated shelf-life testing method. Pasteurized orange juice samples were stored at 10, 20, 30 and 40 °C and tested for acceptance or rejection at consecutive time points using a reverse product preservation plan. From a total of 504 samples and the participation of 72 consumers, 155 failures and 109 passes were recorded. Due to the large number of samples (28) selected for testing from each consumer, a balanced incomplete block design was adopted to facilitate a very small number for evaluation (7 for each consumer) with minimal reduction in the reliability of the experiment and results. As expected, there was a gradual increasing trend of rejections at all temperature levels by time. Also, it was observed that pasteurized orange juices tended to be rejected by consumers at approximately 3 days before they would be microbiologically unsafe for consumption. Also, it was shown from the risk rate trend that higher storage temperatures are linked with very early perceptiveness of sensory defects during nearly the first 100 h of preservation. The present work constitutes essentially a very useful tool for the industry because it enables the proper adjustment of shelf life including both sensory and microbiological parameters, which results in fewer returned and deteriorated products.

Keywords: sensory shelf life; consumer acceptability; accelerated shelf-life conditions; balanced incomplete block designs

1. Introduction

Orange (*Citrus cinensis*) juice is undoubtedly one of the most consumed products worldwide and this is mainly due to its nutritional value, as it is a source of vitamin C and other antioxidant compounds [1], but is also due to its distinguishing sensory characteristics, which can be attributed to the presence of a variety of aldehydes, esters, ketones, terpenes and alcohols. Plus, orange juice is the most widespread among fruit juices, representing almost 45% of the market worldwide [2].

Due to its global distribution, the need has emerged for increasing shelf life of orange juice, because fresh squeezed orange juice was estimated to have a shelf life of ~7 days due to microbiological and physicochemical deterioration [3]. For that reason, thermal pasteurization of citrus juices, which mainly focuses on enzyme inactivation and deteriorative microflora, is the most applied technique globally [4]. Although with thermal pasteurization of citrus juices safety is guaranteed, there is a significant impact on the sensory characteristics of the end product and more specifically on color, odor and taste, so proper selection of time and temperature combination in thermal processing is important [5].

Although it is very easy nowadays to evaluate with laboratory techniques how compromised is the microbiological quality of a product, evaluation of sensory deterioration is still challenging, but the correlation between microbiological and sensory quality, and more specifically safety and sensory shelf life, is extremely important [6,7]. Accelerated shelf-life testing (ASLT) is a reliable tool for establishing a product's shelf life without the need for extensive time or tools, using temperature as an acceleration factor [8–10].

When accelerated shelf-life tests are used, products are typically stored under a variety of temperatures, and at distinctive time periods, measures are taken in order to evaluate possible defects of the products. Different types of pasteurized fruit juices, including orange, were evaluated sensorially stored at different temperatures in order to specify the threshold where the defect appears and the nature of it [11–16].

Balanced incomplete block designs (BIBDs) are essentially innovative short-term sensory test techniques that include untrained assessors and provide reliable results. More specifically, BIBDs are recommended when a large number of samples under testing induce sensory fatigue, leading to compromised results. For the use of BIBDs, a number of prerequisites need to be met such as: (1) at each sensory session, the panelists only evaluate a fraction (k) of the total number of samples (t), (2) all samples are evaluated the same number of times (r), (3) the number of encounters of a similar pair of samples is the same for every other pair, defined as (λ), (4) the number of blocks (b) each including a different fraction of samples, which is also equivalent to the number of participants, needed to run an overall cycle of the experiment [17–19].

So far, however, there has not been a large-scale sensory experiment using untrained consumers, which is the target group of the market, for the evaluation of the sensory shelf life of orange juice stored under different temperatures, and also the correlation between sensory and microbiological shelf life has not been investigated. Thus, the present study aims mainly to assess the consumer acceptance rate of pasteurized orange juice stored under different temperature levels and secondly to investigate whether rejections of orange juices based on sensory defects perceived by consumers appear within the microbiological safety limits. Due to the innovative character of the present work and the methodology used, literature regarding its content is scarce.

2. Materials and Methods

2.1. Materials

Commercial pasteurized orange juice was used in the present study and was manufactured at a local juice production factory in 500 mL PET (Polyethylene terephthalate) bottles. Directly after production, pasteurized orange juice bottles were stored under different temperatures in the company's industrial refrigerators.

2.2. Sensory Evaluation

Based on our previous results regarding orange juice production [20], the company extracted the results of the microbiological shelf life of the product at 10, 20, 30 and 40 °C temperature levels for further use and chose to apply the exponential model, resulting in the following microbiological shelf life: 363, 159, 86 and 74 h or 15.1, 6.6, 3.3 and 3.1 days (10, 20, 30, 40 °C). It is reiterated that the microbiological shelf life of the fruit juices was based on measurements of the microbial load up to the 1000 cfu/mL APC (Aerobic Plate Count) limit for safe consumption and minimum risk purposes (20).

A large-scale experiment on the sensory evaluation of orange juices was therefore organized to answer the question whether consumers were able to intuitively assess the potential organoleptic degradation of products at different temperature levels (always within the 1000 cfu/mL APC safety limit) and at which specific preservation intervals this manifests itself. Based on the results of the previous work [20] concerning microbiological deterioration and color degradation, values of pH and colorimetric b* were statistically significant. Thus, consumers were asked to accept or reject each sample according to sensing of sourness and dark orange color.

To this end, a large-scale sensory experiment of fruit juice testing was organized with the participation of members of the company's staff, numbering more than 100 employees of different specialties, the majority of whom belong to the category of typical consumers. This food industry has an old hall for performing sensory tests on its products. The hall has an oval table surrounded by 21 workstations. Six equally spaced panelists participated each time, and there were randomly ordered seven samples in plastic beakers each identified with a three-digit number, plus one beaker filled with water and another empty for other use. The reverse product preservation plan [21] was chosen, through which samples preserved at different initial times were finally synchronized and all appear ready for testing at time 0. Finally, 7 equivalent time intervals were selected at each temperature level, the ranges of which were adjusted according to the shelf life of the orange juice.

The inverse time sampling is a reverse preservation plan for the products, whereby samples are stored at successive predetermined times with countdown to day 0. A full comparative sensory check is therefore carried out at time 0 with the inclusion of all preserved samples at the previous times, as shown in the indicative Scheme 1 below:

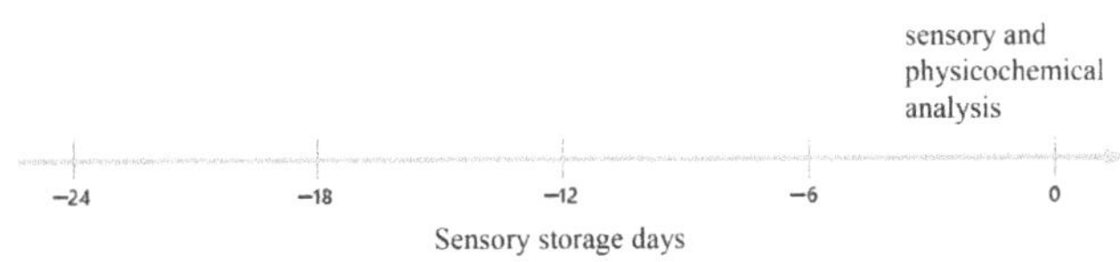

Scheme 1. Inverse preservation plan process stages.

For example, in the reverse time plot, the first sample is preserved for 24 days, the second sample for 18 days and so on and at the end of the preservation period, i.e., at time 0, all samples are ready for a comparative sensory test and/or physico-chemical measurements. The sensory shelf life is based on the decision by the consumer to reject or not to reject a product and is statistically investigated by applying survival analysis, i.e., estimating the lifetime until the event of interest (rejection) takes place. The consumer tests samples of different shelf lives in experimental time-reversal designs and decides whether or not to buy some of them (product acceptance/rejection). The minimum number of consumers participating in similar sensory tests is between 50 and 200. The recording of the responses includes left and right censored data or even intervals. The non-parametric survival analysis is calculated using the maximum likelihood equation, which describes the joint probability of the observations of a sample of n consumers:

$$L = \prod_{i \in R} S(r_i) * \prod_{i \in L} (1 - S(l_i)) * \prod_{i \in I} (S(l_i - S(r_i))$$

where L and R are, respectively, the groups of left and right censored items and I is the interval group of observations. This equation without the left-censored items and with exact recording times is applied in this research.

Thus, on the basis of the final shelf life that was recorded per temperature, up to the limit of 1000 APC (cfu/mL) units, the following calendar table was established (Table 1):

Table 1. Time intervals for orange juice sampling at the corresponding storage temperatures (h in blue color).

Hours	Temperature (°C)						
	10		**20**		**30**		**40**
1	52	8	24	15	12	22	10
2	104	9	48	16	24	23	20
3	156	10	72	17	36	24	30
4	208	11	96	18	48	25	40
5	260	12	120	19	60	26	50
6	312	13	144	20	72	27	60
7	364	14	168	21	84	28	70

For the 10 °C level, the time intervals were divided into 52 h, for the 20 °C level, they were divided into 24 h and for the last two, 30 and 40 °C, they were divided into 12 and 10 h, respectively. For example, sample 19 refers to 60 h of preservation at 30 °C, and the sample 13 refers to 144 h at 20 °C. Because in this reverse sensory scheme 28 samples (7 × 4) were synchronized to time 0 to be tested by each panelist, this number was considered extremely high and a Balanced Incomplete Block (BIB) design was adopted (plan 11.39, p. 482, Cochran and Cox 1957) [18], shown in Table 2 and including the following characteristics:

Table 2. A balanced incomplete block design adapted for the sensory requirements of each consumer (b).

b	t = 28, k = 7, b = 36, n = 9, λ = 2 Treatments (t)						
1	4	7	8	9	14	23	28
2	1	5	9	10	11	15	24
3	6	8	13	15	16	18	21
4	7	12	13	17	22	24	25
5	4	10	16	17	20	26	27
6	2	11	18	19	22	26	28
7	1	3	6	12	19	23	27
8	2	3	5	14	20	21	25
9	1	2	8	10	12	16	25
10	2	3	6	9	11	13	17
11	4	5	12	13	15	19	26
12	3	7	16	18	19	20	24
13	6	10	14	21	22	24	26
14	11	15	20	22	23	25	27
15	1	5	17	18	18	27	28
16	2	7	9	12	21	26	27

Table 2. *Cont.*

b	t = 28, k = 7, b = 36, n = 9, λ = 2 Treatments (t)						
17	3	8	10	13	22	27	28
18	6	12	14	15	17	20	28
19	4	8	11	17	19	21	25
20	1	13	14	18	23	25	26
21	2	4	5	6	16	22	23
22	3	4	10	11	12	14	18
23	1	9	14	16	17	19	22
24	1	2	4	13	20	24	28
25	3	5	8	17	23	24	26
26	5	6	7	10	19	25	28
27	1	6	7	8	11	20	26
28	9	10	13	19	20	21	23
29	2	8	14	15	19	25	27
30	3	9	15	16	25	26	28
31	5	8	9	12	18	20	22
32	11	12	16	21	23	24	28
33	1	3	4	7	15	21	22
34	5	7	11	13	14	16	27
35	4	6	9	18	24	25	27
36	2	7	10	15	17	18	23

Each consumer tastes only 7 different samples, randomly selected using the JMP statistical program, and in total 36 individuals are required, but to enhance the reliability of the results, the experiment was performed twice. The experimental BIB design includes t = 28 treatments (samples), k = 7 treatments per panelist, b = 36 panelists, n = 9 individual samples per combined level (time by temperature) of the calendar table and λ = 2 similar paired encounters (Table 2). Each consumer was given a simple questionnaire with 7 different 3-digit codes corresponding to 7 treatments in order after randomization, asking whether after each individual trial he would buy the particular juice, i.e., whether the sample is acceptable (Y) or rejected (N). Before the start of the trials, the consumer was also offered a sample of fresh juice, and if rejected, the sensory procedure for the consumer was immediately terminated (invalid). For the execution of the experiment, only the staff members who were used to consuming at least one fruit juice per three-day period were chosen for the experiment.

2.3. Statistical Analysis

The Kaplan–Meier method was used for the sensory test, while the log-rank and Kruskal–Wallis tests of homogeneity were used to examine non-parametric differences between the samples. All statistical analyses were performed via JMP 17.2 (Statistical

Discovery LLC., Cary, NC, USA) software. Probability values less than the reference level of 0.05 were considered statistically significant.

3. Results

Sensory Evaluation of Orange Juice at Different Storage Temperature Levels and Time

The results of the sensory analysis of the 72 consumers were recorded and presented in Table 3, in which the acceptance (pass) or rejection (failure) of the samples at different time intervals and at each temperature level (different coloring) is indicated in each row. First, it can be observed that when a storage period at a temperature level included one or more passes, only the last observed decision was recorded. For example, in the second row and 10 °C time period, only the acceptance of >260 h was marked, and for the 40 °C time period, only the acceptance of >30 h was noted. Secondly, when a time interval included both passes and failures, only that when the first failure occurred was taken into account for the estimation of the shelf life of the orange juices. For example, in row 43 and at the 10 °C temperature level, only the 156-h rejection was recorded.

The second run of the experiment facilitated the recording of 18 (9 × 2) replicates per treatment, and their allocation to the pass and failure frequencies is described at the end of the worksheet. As expected, there was a gradual increasing trend of rejections at all temperature levels by time and particularly a two-fold rise in rejections (13) was revealed in the last 364-h time interval at 10 °C. This observation was expected, because it is known also in literature that liking and desirability of orange juices declines with increasing storage times or temperature [14,15].

The failure rate of the juices as a function of lifetime is illustrated by the survival curves (Figure 1), showing steep decreasing trends of survival at temperatures 30 °C and 40 °C, a mild decreasing trend at 20 °C and a very slow rate at 10 °C. The curves represent the rate of the survival change by applying the lognormal distribution together with 95% confidence intervals.

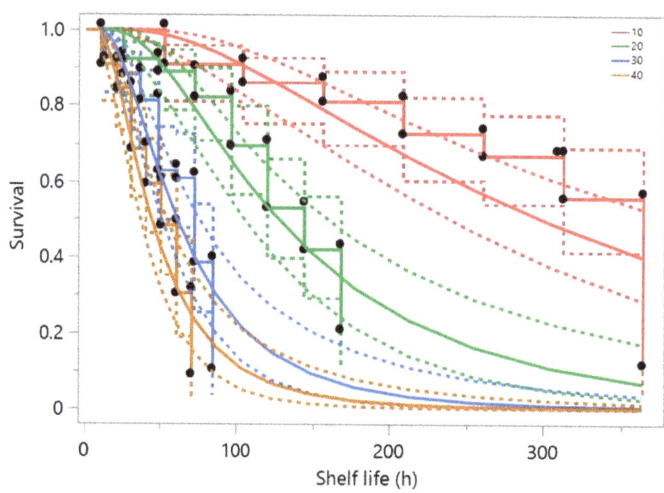

Figure 1. Survival curves with 95% confidence intervals (dashed lines) of pasteurized orange juice stored at different temperature levels.

Table 3. Acceptances (Y) and rejections (N) of orange juices at different storage times and temperature levels.

	10							20							30							40							10	20	30	40
Time (h)	52	104	156	208	260	312	364	24	48	72	96	120	144	168	12	24	36	48	60	72	84	10	20	30	40	50	60	70				
1				Y			Y	Y	Y					Y									Y					N	>364	>168		70
2	Y				Y				Y	Y	N				N									Y					>260	96	12	>30
3						N		Y					N		Y	Y		Y			N								312	144	84	
4							Y					Y	Y				N					Y		Y	Y				>364	>144	36	>40
5				N						N						Y	Y			Y						Y	N		208	72	>72	60
6		N									Y							N	Y			N				Y		N	104	>96	48	10
7	N		N			Y						N		Y					Y				N				Y		52	120	>60	20
8		Y	Y		Y															Y	N				Y				>260	>168	84	>40
9	N	Y						N		Y		Y				Y									Y				52	24	>24	>40
10		Y	Y			Y			Y		N		Y				Y												>312	96	>36	
11				Y	Y							Y	Y		N				Y							Y			>260	>144	12	>50
12			Y													Y		N	Y	N				Y					364		48	>30
13						Y				N				N							Y	Y		N		N			>312	72	>84	30
14											Y				N					N		Y	N		Y		Y			>96	12	20
15	Y				N												Y	N			N						N	N	260	>208	48	60
16		Y							Y			N									N					Y	Y		364	120	84	>60
17			Y					Y		Y			Y									Y					N	N	>208	>144		60
18						Y						Y		N	Y		N			N								N	>312	168	36	70
19				Y				N			N						Y		Y		N				N				>308	24	84	40
20	Y												Y	N				Y					Y		N	N			>52	168	>48	40
21		Y		Y	Y	N										Y						N	N						312	52	>24	10
22			Y	Y						Y	Y	Y		Y				Y											>208	>168	>48	
23	N								N					Y		N	Y		N			N							52	48	24	10
24	N	N		Y									N							Y				Y				N	52	144	>72	70
25			Y		Y			N									Y						Y	Y		Y			>260	24	>36	>50
26					Y	Y	N			Y									Y						Y				364	>72	>60	>70
27	Y					Y	N	Y			N									N						N			364	96	72	50
28									Y	Y			Y						Y	N	N		Y							>144	72	>20

Table 3. *Cont.*

	10							20							30							40							10	20	30	40
Time (h)	**52**	**104**	**156**	**208**	**260**	**312**	**364**	**24**	**48**	**72**	**96**	**120**	**144**	**168**	**12**	**24**	**36**	**48**	**60**	**72**	**84**	**10**	**20**	**30**	**40**	**50**	**60**	**70**				
29		Y						Y						Y	Y				Y					N			Y		>104	>168	>60	30
30			Y						Y						Y	Y									Y	Y		Y	>156	>48	>24	>70
31					Y			Y	Y			N						N		Y		N							>260	120	48	10
32											Y	Y									N		Y	N				N		>120	24	30
33	Y		Y	N			N								Y						Y	Y							208		>84	>10
34					Y		Y				Y		N	Y		Y											Y		>364	144	>24	>60
35				Y		N			Y									N						Y	N		Y		312	>48	48	40
36		Y					N			Y					Y		Y	N					Y						364	>72	48	>20
37				N			N	Y	Y					N									Y					Y	208	168		>20
38	N				N				Y	Y	Y				N									N					52	>96	12	30
39						Y		Y					Y		N	N		Y			N								>312	>144	12	
40							N					N	Y				Y					Y		N	Y				364	120	>36	30
41				N						Y						Y	N			Y						N	Y		208	>72	36	50
42		N									N							N	N			Y				Y		N	104	96	48	70
43	Y		N			N						N							Y				N				Y		156	120	>60	20
44		Y	Y		N									Y						N	Y				Y				260	>168	72	>40
45	N	Y						Y		N		Y				Y									N				52	725	>24	40
46		N	N			N			Y		N		Y				Y												104	96	>36	
47				Y	Y							N	N		Y				N							N			>260	120	60	50
48			Y													N		N	N	Y				N					364		24	30
49										Y				N							N	Y		N		N			312	168	84	30
50											N				Y					Y		N	Y		N		N			96	>72	10
51	Y				Y												Y	N			N						N	N	>260		48	60
52		Y					N		Y			N									N					N	N		364	120	84	50
53			N					N		Y			N									Y					N	N	156	24		60
54						Y						Y		Y	Y		Y			N								N	>312	>168	72	70
55				Y				Y			N						Y		Y		N				Y				>208	96	84	>40
56	Y												Y	N				N					Y		Y	N			>52	168	48	50
57		Y		Y	Y	Y										Y						Y	Y						>312		>24	>20
58			Y	Y						Y	Y	Y		Y				N											>208	>168	48	

Table 3. *Cont.*

	10							20							30							40							10	20	30	40
Time (h)	**52**	**104**	**156**	**208**	**260**	**312**	**364**	**24**	**48**	**72**	**96**	**120**	**144**	**168**	**12**	**24**	**36**	**48**	**60**	**72**	**84**	**10**	**20**	**30**	**40**	**50**	**60**	**70**				
59	Y								Y					N		Y	Y		Y			Y							>52	168	>60	>10
60	Y	Y		N									N							N				Y				N	208	144	72	70
61			Y		Y			N									N						Y	N		N			>260	24	36	30
62					N	N	Y			N									Y						Y				260	72	>60	>70
63	Y					Y	N	Y			Y		Y							N						Y			364	>96	72	>50
64									N	Y									Y	N	N		Y							48	72	>20
65		Y						Y						N	Y				Y					Y			N		>104	168	>60	60
66			N						Y						Y	Y									N	N		N	156	>48	>24	40
67					Y			Y	Y			N						Y		N		Y							>260	120	72	>10
68											Y	Y									N		Y	Y				N		>120	84	70
69	Y		Y	Y			N								Y						Y								364		>84	10
70					Y		Y				Y		N	N		Y											N		>260	144	>24	60
71				Y		Y			Y									Y						N	Y		N		>208	>48	>48	30
72		Y					N			Y					Y		Y	Y					N						364	>72	>48	20
NO	6	4	5	5	4	7	13	5	2	4	8	8	7	9	5	4	4	11	4	11	14	6	5	9	6	10	10	14				
YES	12	14	13	13	14	11	5	13	16	14	10	10	11	9	13	14	14	7	14	7	4	12	13	9	12	8	8	4				
	18	18	18	18	18	18	18	18	18	18	18	18	18	18	18	18	18	18	18	18	18	18	18	18	18	18	18	18				

In summary, from a total of 504 samples and with the participation of 72 consumers, 155 failures and 109 passes were recorded, with the number of rejections increasing with temperature rise (Table 4). Using the Kaplan–Meier estimator, the sensory shelf life of orange juices is 287.9 h (11.9 d) at 10 °C, 126.8 h (5.3 d) at 20 °C and much lower at higher temperature levels of 30 and 40 °C (63 and 48.2 h, 2.6 and 2 d, respectively) (Table 4). In the study of Klimczak et al. (2011) [11], orange juice was stored for 12 months at 18, 28 and 38 °C, and while there was no obvious deterioration during preservation at 18 degrees after 1 year, undesirable organoleptic characteristics were observed at the highest preservation temperature levels after only 6 months. The very high shelf life of the product used in the study above was maybe due to processing parameters. In another study of Oliveira et al. (2011) [12], where pasteurized mango juices were stored at 25, 35 and 45 °C for 120 days, it was reported that sensory aroma and color were kept stable at lower temperature levels throughout the whole storage time, but on the contrary, at higher temperature levels, consumers were able to detect deterioration even from the first days, which was also observed in the present study. The rise of rejections, which is more apparent at high temperature levels, was due to the degradation of thermosensitive nutrients such as vitamin C and others that lead to degradation of sensory characteristics such as aroma, flavor and taste [22–24]. The data in Table 5, which presents the comparison between the sensory and the microbiological shelf life, show that the sensory shelf life corresponds to 12, 5.3, 2.6 and 2.0 days (Temperature levels: 10, 20, 30, 40 °C), all of which were lower compared to the microbiological shelf life: 15.1, 6.6, 3.6 and 3.1 days (exponential model), and 16.4, 7.2, 4.5 and 3.2 days (polynomial model) [17]. This deviation was mainly due to the more rapid degradation kinetics of enzymatic reactions than that of microbiological components. This means basically that the juice maybe presented unwanted sensory characteristics within its microbiological shelf life. This is very important for the industry because if they take into account both microbiological and sensory shelf life while adjusting the expiration date of the pasteurized orange juice, they will minimize returned products. More specifically, off-flavors and unwanted aromas of orange juice are mainly attributed to the generation of 4-vinylguaiacol (4VG) after prolonged storage at higher temperatures [25]. According to the work of Kopuncova et al. (2022) [26], which assessed the most prominent types of off-flavor and off-taste chemical compounds that were generated during long-term storage of orange juice at high temperatures, four compounds and more specifically dimethyl sulphide, 4-vinylguaiacol, α-terpineol and furaneol were identified.

Table 4. Survival time (h) at different temperature levels using the Kaplan–Meier estimator.

Temperature Level (°C)	Failures	Passes	Median Time (h)	Days	SE
10	35	31	287.9	12	13.8
20	38	26	126.8	5.3	6.0
30	39	29	63.0	2.6	3.0
40	43	23	48.2	2.0	2.7
Combined	155	109	168.9		9.6

Table 6 provides information on the shelf life of juices and more specifically on the median shelf life (h) per temperature level, including also the 95% confidence intervals and the quartiles of the median. The log-rank and Kruskal–Wallis test of homogeneity between temperature levels showed strong statistical significance ($p < 0.0001$) (Table 7), indicating different median survival times, and judging by the size of the overlap of both the quartiles and the confidence limits, a survival decrease at the levels: 10 > 20 > 30 = 40 is apparent (in the last two temperature levels, the confidence limits of survival overlap significantly).

Table 5. Comparison of sensory with microbial shelf life of pasteurized orange juice.

Temperature Level (°C)	Survival Time (Days)		
	Sensory Shelf Life	Microbiological Shelf Life	
		Exponential Model	Polynomial Model
10	12	15.1	16.4
20	5.3	6.6	7.2
30	2.6	3.6	4.5
40	2.0	3.1	3.2

Table 6. Sensory shelf life of pasteurized orange juice at different temperature levels.

Temperature Level (°C)	Median Time (h)	Days	−95%	+95%	25% Failures	75% Failures
10	364	15.2	312	364	208	364
20	144	6.0	120	168	96	168
30	72	3.0	48	84	48	84
40	50	2.1	40	60	30	70
Combined	120		84	144	52	312

Table 7. Homogeneity test of survival data between temperature levels.

Homogeneity	$\chi 2$	DF	p
Log-rank	153.12	3	<0.0001
Kruskal–Wallis	110.82	3	<0.0001

The analytical survival rate of orange juice data (Table 8) indicated the shelf life and the survival rate per temperature level. The term "pass" and "failure" correspond to the "acceptance" or "rejection" of a sample. So, depending on the rejection rate that a company considers acceptable, the corresponding shelf life can be calculated. Thus, if the firm considers the minimum acceptance rate for its product to be around 50% (e.g., 50% of the consumers will reject the product) the predicted sensory shelf life (h) is: 312 h at 10 °C (44.2% failure), 120 h at 30 °C (46.83% failure), 60 h at 30 °C (39.2%) and 50 h at 40 °C (51.6%). In the study of Buve et al. (2016) [16], where strawberry juice was stored at different temperature levels, it was shown that for an acceptance rate 50% the predicted sensory shelf life (h) was ~2016, ~1344, ~672, ~336 for 20, 28, 38 and 42 °C. Although direct comparison with the results of the present study was not adequate due to the differences between the products, it was observed that in both experiments increasing storage temperature reduces by half the time needed for the consumers to perceive sensory defects. Similarly, for a more stringent failure rate of <20% or survival rate > 80% (only 20% of the consumers will reject the product), the sensory shelf life varied as follows: 156 h at 10 °C (19% failure), 72 h at 20 °C (18% failure), 36 h at 30 °C (18.6% failure) and 20 h at 40 °C (15.5% failure). In other words, it seems that preservation of the orange juice at 10 °C facilitated an adequate sensory and microbiologically safe shelf life. However, it is very useful and crucial that these data on the sensory failure (rejection) rate assessment of products be linked with objective instrumental measurement [27].

Table 8. Analytical survival rate of pasteurized orange juice shelf life (h) at each temperature level.

Temperature Level (°C)	Shelf Life (h)	Survival	Failure	SE	Number Failed	Number Passed	at Risk
10	0	1	0	0	0	0	66
	52	0.91	0.09	0.04	6	3	66
	104	0.86	0.14	0.04	3	2	57
	156	0.81	0.19	0.05	3	1	52
	208	0.73	0.27	0.06	5	5	48
	260	0.67	0.33	0.06	3	10	38
	308	0.67	0.33	0.06	0	1	25
	312	0.56	0.44	0.07	4	6	24
	364	0.12	0.88	0.06	11	3	14
20	0	1	0	0	0	0	64
	24	0.92	0.08	0.03	5	0	64
	48	0.89	0.11	0.04	2	4	59
	72	0.82	0.18	0.05	4	4	53
	96	0.7	0.3	0.06	7	4	45
	120	0.53	0.47	0.07	8	2	34
	144	0.42	0.58	0.07	5	5	24
	168	0.21	0.79	0.07	7	7	14
30	0	1	0	0	0	0	68
	12	0.93	0.07	0.03	5	0	68
	24	0.88	0.12	0.04	3	8	63
	36	0.81	0.19	0.05	4	4	52
	48	0.63	0.37	0.06	10	4	44
	60	0.61	0.39	0.07	1	7	30
	72	0.39	0.61	0.07	8	3	22
	84	0.11	0.89	0.06	8	3	11
40	0	1	0	0	0	0	66
	10	0.91	0.09	0.04	6	3	66
	20	0.85	0.15	0.05	4	5	57
	30	0.69	0.31	0.06	9	2	48
	40	0.59	0.41	0.06	5	5	37
	50	0.48	0.52	0.07	5	3	27
	60	0.31	0.69	0.07	7	2	19
	70	0.09	0.91	0.05	7	3	10

As an alternative approach, various regression models were attempted (exponential, weibull, lognormal), which revealed lognormal as the best fit performed at each temperature level (Figure 2).

Although the study points were not scattered equally on either side of the regression lines at all temperature levels (Figure 2), the estimates of the average shelf life in the table of parameters of the lognormal distribution yielded comparatively reliable results. In particular, the lognormal values of the survival times (h) were very close to those of the exponential model and thus indicated a second reliable model for fitting and predicting the sensory shelf-life data (Table 9).

After applying the log-normal transformation, the fitted distributions of the analytical characteristics of survival time were obtained (Figure 3): At 80% survival, the shelf-life curves of the two highest temperatures started to differentiate remarkably with a 10-h difference from 40 to 30 °C, which doubled at 50% survival. Furthermore, at 20 and 10 °C and at 80% survival rate, the difference was significant (approximately 80 and 180 h, respectively) and became very large at 50% (approximately 175 and 300 h, respectively) (Figure 3a). This is in agreement with the results found by Buve et al. (2016) [16], where it was observed that acceptability rates do not have the same extrapolation with temperature levels. In the density distribution graph (Figure 3b), the peak of the 40 °C density curve was observed at 28 h, and, respectively, at 38, 76 and 150 h at the other temperature levels, showing the strong effect of heat stress on sensory shelf life, the magnitude of which was determined by the area occupied by the distribution spreads in the shelf life, which were very limited at high temperatures (30, 40 °C) and extended at lower temperatures (Figure 3b). The risk rate showed initially a steep upward trend at higher temperatures and peaked in less than 80 h, in contrast to the lower temperatures that showed a lower rate, which was extremely low at 10 °C (Figure 3c).

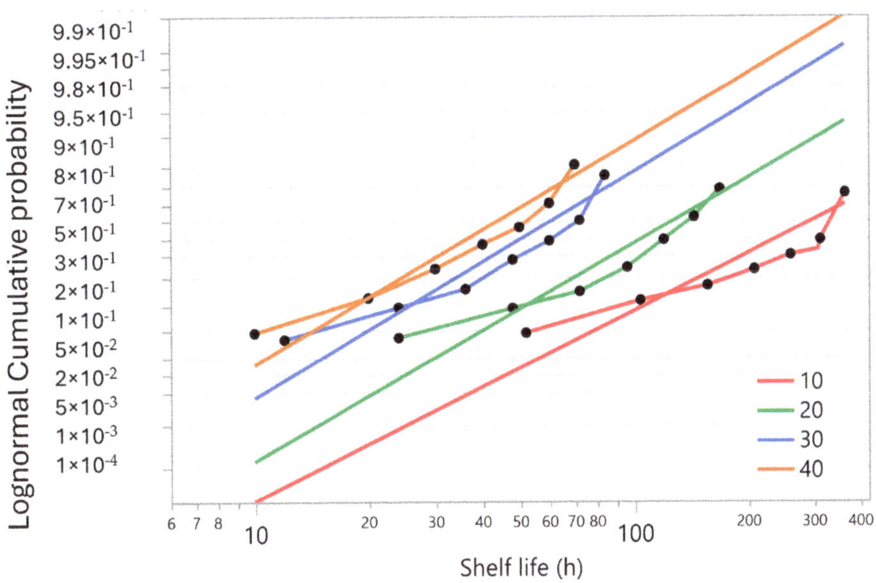

Figure 2. Cumulative failure probability versus shelf life of pasteurized orange juice at each storage temperature level of the lognormal transformation.

Table 9. Lognormal distribution parameters of pasteurized orange juice shelf life stored at different temperature levels.

Temperature (°C)	Parameter	Estimate	H	Days	Lower 95%	Upper 95%	Number Failed
10	μ	5.7	298	12.4	5.48	5.96	35
10	σ	0.8	298	12.4	0.64	1.04	35
20	μ	4.85	127.5	5.3	4.66	5.07	38
20	σ	0.7	127.5	5.3	0.57	0.9	38
30	μ	4.11	60.8	2.5	3.93	4.32	39
30	σ	0.67	60.8	2.5	0.55	0.86	39
40	μ	3.79	0	1.8	3.6	3.98	43
40	σ	0.68	0	1.8	0.56	0.86	43

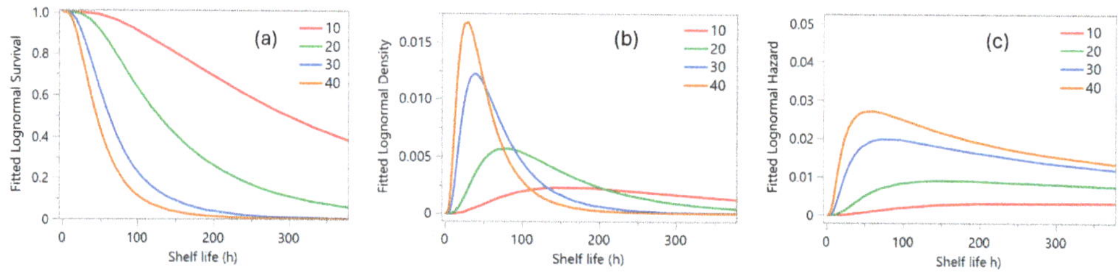

Figure 3. Fitted distributions of the analytical characteristics of survival time using lognormal transformation at each temperature level (**a**) survival curves as a function of shelf life (h), (**b**) density curves as a function of shelf life (h), (**c**) hazard curves as a function of shelf life (h).

4. Conclusions

In the present study, an accelerated shelf-life testing method with temperature as an acceleration factor was used to evaluate the sensorial acceptability of orange juices stored at different temperature levels for prolonged time periods. The panel was comprised of untrained consumers, and data revealed that they were able to perceive sensory defects very early in the preservation period, which was related with the storage temperature of juices. Increasing temperature by 10 °C led generally to the reduction by half of the sensory shelf life provided that acceptance rate was around ~50%. Moreover, if considering the results from our previous study [20] dealing with the microbiological shelf life of pasteurized orange juices, it was observed that pasteurized orange juices tended to be sensorially rejected by consumers ~3 days before they would be microbiologically unstable for consumption. These results indicate that it is very important, much more for microbiologically safe products like orange juices, when manufacturing companies and government agencies try to establish shelf-life to take into account the sensory shelf life in order to not have losses of products and complaints about overall quality.

Author Contributions: Conceptualization, D.P. and A.K.; methodology, D.P. and A.K.; software, D.P. and T.F.; validation, D.P., A.K. and T.F.; formal analysis, D.P. and T.F.; investigation, D.P. and A.K.; resources, A.K.; data curation, D.P. and T.F.; writing—original draft preparation, T.F. and D.P.; writing—review and editing, T.F. and D.P.; visualization, D.P. and A.K.; supervision, D.P.; project administration, D.P.; funding acquisition, D.P., A.K. and T.F. All authors have read and agreed to the published version of the manuscript.

Funding: This research received no external funding.

Institutional Review Board Statement: Not applicable.

Informed Consent Statement: Not applicable.

Data Availability Statement: The original contributions presented in the study are included in the article; further inquiries can be directed to the corresponding author.

Conflicts of Interest: The authors declare no conflicts of interest.

References

1. Dhenge, R.; Alinovi, M.; Lolli, V.; Aldini, A.; Rinaldi, M. Evaluation of quality parameters of orange juice stabilized by two thermal treatments (helical heat exchanger and ohmic heating) and non-thermal (high-pressure processing). *Food Control* **2022**, *141*, 109–150. [CrossRef]
2. Pan, X.; Bi, S.; Lao, F.; Wu, J. Factors affecting aroma compounds in orange juice and their sensory perception: A review. *Food Res. Int.* **2021**, *169*, 112835. [CrossRef]

3. Rodrigues, P.V.; Vieira, D.M.; Martins, P.C.; Martins, V.G.; Castro, M.C.R.; Machado, A.V. Evaluation of Active LDPE Films for Packaging of Fresh Orange Juice. *Polymers* **2023**, *15*, 50. [CrossRef] [PubMed]
4. Yıldız, D.; Benzer Gürel, D.; Çağındı, O.; Kayaardı, S. Heat treatment and microwave applications on homemade sour cherry juice: The effect on anthocyanin content and some physicochemical properties. *Curr. Plant Biol.* **2023**, *29*, 100242. [CrossRef]
5. Kumar, V.; Kohli, D.; Naik, B.; Ratore, A.; Kumar Gupta, A.; Masood Khan, J.; Irfan, M.; Preet, M.S.; Chatterjee, N.; Rustagi, S. Effect of heat treatment on the quality of citrus juices. *J. King Saud Univ. Sci.* **2023**, *35*, 102819. [CrossRef]
6. Hough, G.; Garitta, L. Methodology for sensory shelf-life estimation: A review. *J. Sens. Stud.* **2012**, *27*, 137–147. [CrossRef]
7. Giménez, A.; Ares, F.; Ares, G. Sensory shelf-life estimation: A review of current methodological approaches. *Food Res. Int.* **2012**, *49*, 311–325. [CrossRef]
8. Haouet, M.N.; Tommasino, M.; Mercuri, M.L.; Benedetti, F.; Bella, S.D.; Framboas, M.; Pelli, S.; Altissimi, M.S. Experimental accelerated shelf life determination of a ready-to-eat processed food. *Ital. J. Food Saf.* **2019**, *11*, 6919. [CrossRef] [PubMed]
9. Pedro, A.M.; Ferreira, M.M. Multivariate accelerated shelf-life testing: A novel approach for determining the shelf-life of foods. *J. Chemom.* **2006**, *20*, 76–83. [CrossRef]
10. Nurhayati, R.; Susanto, A.; Khasanah, Y. Shelf life prediction for canned gudeg using accelerated shelf life testing (ASLT) based on Arrhenius method. *IOP Conf. Ser. Mater. Sci. Eng.* **2019**, *193*, 012025. [CrossRef]
11. Klimczak, I.; Małecka, M. Evaluation of sensory profile and p-vinylguaiacol (pvg) content in orange juices during storage at different temperature. *J. Food Qual.* **2011**, *34*, 30–39. [CrossRef]
12. Oliveira, A.; Ramos, A.M.; Minim, V.P.R.; Chaves, J.B.P. Sensory stability of whole mango juice: Influence of temperature and storage time. *Food Sci. Technol.* **2012**, *32*, 819–825. [CrossRef]
13. Kishore, K.; Pathak, K.A.; Shukla, R.; Bharali, R. Effect of storage temperature on physico-chemical and sensory attributes of purple passion fruit (*Passiflora edulis* Sims). *J. Food Sci. Technol.* **2011**, *48*, 484–488. [CrossRef]
14. Song, Q.; Rune, C.J.B.; Thybo, A.K.; Clausen, M.P.; Orlien, V.; Giacalone, D. Sensory quality and consumer perception of high pressure processed orange juice and apple juice. *Lwt* **2023**, *173*, 114303. [CrossRef]
15. Polydera, A.C.; Stoforos, N.G.; Taoukis, P.S. Quality degradation kinetics of pasteurised and high pressure processed fresh Navel orange juice: Nutritional parameters and shelf life. *Innov. Food Sci. Emerg. Technol.* **2005**, *6*, 1–9. [CrossRef]
16. Buvé, C.; Van Bedts, T.; Haenen, A.; Kebede, B.; Braekers, R.; Hendrickx, M.; Van Loey, A.; Grauwet, T. Shelf-life dating of shelf-stable strawberry juice based on survival analysis of consumer acceptance information. *J. Sci. Food Agric.* **2018**, *98*, 3437–3445. [CrossRef] [PubMed]
17. dos Santos Navarro, R.D.C.; Minim, V.P.R.; da Silva, A.N.; Simiqueli, A.A.; Della Lucia, S.M.; Minim, L.A. Balanced incomplete block design: An alternative for data collection in the optimized descriptive profile. *Food Res. Int.* **2014**, *64*, 289–297. [CrossRef]
18. Cochran, W.G.; Cox, G.M. *Experimental Design*, 2nd ed.; John Wiley and Sons: New York, NY, USA, 1957; p. 615.
19. *ISO 29842:2024*; Sensory analysis—Methodology—Balanced incomplete block designs. ISO: Geneva, Switzerland, 2024.
20. Frangopoulos, T.; Koliouskas, A.; Petridis, D. The Effect of Accelerated Storage Temperature Conditions on the Shelf Life of Pasteurized Orange Juice Based on Microbiological, Physicochemical, and Color Attributes. *Appl. Sci.* **2024**, *14*, 10870. [CrossRef]
21. Surkan, S.; Albani, O.; Ramallo, L. Influence of storage conditions on sensory shelf life of yerba mate. *J. Food Qual.* **2009**, *32*, 58–72. [CrossRef]
22. Zhang, J.; Han, H.; Xia, J.; Gao, M. Degradation kinetics of vitamin C in orange and orange juice during storage. *Adv. J. Food Sci. Technol.* **2016**, *12*, 555–561. [CrossRef]
23. Burdurlu, H.S.; Koca, N.; Karadeniz, F. Degradation of vitamin C in citrus juice concentrates during storage. *J. Food Eng.* **2006**, *74*, 211–216. [CrossRef]
24. Li, X.; Ren, J.N.; Fan, G.; Pan, S.Y. Changes of aroma compounds and qualities of freshly-squeezed orange juice during storage. *J. Food Sci. Technol.* **2018**, *55*, 4530–4543. [CrossRef] [PubMed]
25. Bauersachs, E.; Walser, V.; Reglitz, K.; Dawid, C.; Steinhaus, M. Peracetic acid residues in orange juice can lead to a 5-vinylguaiacol-induced clove-like off-flavor via Baeyer-Villiger oxidation of hesperidin. *Food Chem.* **2024**, *440*, 138252. [CrossRef]
26. Kopuncová, M.; Sádecká, J.; Kolek, E.; Dimitrov, F.; Tobolková, B.; Durec, J.; Blaško, J. Aroma stability and sensory aspects of commercially produced orange juice: Gas chromatography–olfactometry study. *Chem. Pap.* **2022**, *76*, 4941–4953. [CrossRef]
27. Gámbaro, A.; Fiszman, S.; Giménez, A.; Varela, P.; Salvador, A. Consumer acceptability compared with sensory and instrumental measures of white pan bread: Sensory shelf-life estimation by survival analysis. *J. Food Sci.* **2004**, *69*, S401–S405. [CrossRef]

applied sciences

MDPI

Article

Thermal Behavior and Infrared Absorbance Bands of Citric Acid

Costas Tsioptsias *, Afroditi Panagiotou and Paraskevi Mitlianga *

Laboratory of Food Chemistry and Technology, Department of Chemical Engineering, University of Western Macedonia, 50132 Kozani, Greece; chemeng00161@uowm.gr

* Correspondence: aff00285@uowm.gr or ktsiopts@gmail.com (C.T.); pmitliagka@uowm.gr (P.M.)

Abstract: Citric acid is widely used in the Food and Pharmaceutical Industry. Various issues regarding its thermal behavior and infrared spectrum require clarification. Here, we studied citric acid monohydrate (raw, heated, freeze-dried and recrystallized from D_2O) via Differential Scanning Calorimetry, Thermogravimetric Analysis, Infrared Spectroscopy, and antioxidant capacity assay. Also, we used ab initio Density Functional Theory calculations for further supporting the interpretations of the experimental results. Citric acid monohydrate exhibits desolvation inability and upon heating does not dehydrate but esterifies. Nor by freeze drying can it be dehydrated. The heated sample is not anhydrous, it exhibits melting inability, and any fluidization occurs simultaneously with decomposition. In other words, the interpretations regarding the two endothermic peaks in the DSC curve of citric acid that have been attributed to water evaporation and melting are not correct. The increase in the molecular weight due to esterification is most likely responsible for the increased antioxidant/chelation capacity of the heated sample. We concluded that what we call citric acid monohydrate and anhydrous do not exist in a pure form (in the solid state) and actually are mixtures of different compositions of citric acid, water and a citric acid oligomer that is produced through esterification. The esterification reaction seems to be able to proceed easily under mild heating or even at room temperature. The presence of the ester oligomer and water affect the infrared spectrum of citric acid monohydrate and anhydrous and is responsible for the existence of multiple peaks in the C=O stretching region, which partially overlaps with the water H-O-H bending vibration. The insights presented in this work could be useful for optimizing the design, performance and quality of food and drug products in which citric acid is used.

Keywords: citric acid; melting; decomposition; desolvation; inability; antioxidant; DFT

1. Introduction

Thermal behavior, e.g., the melting point and the thermal stability of materials, is of great importance and is involved in many applications, including material processing, food and drug preparation, etc. Recently, it has been reported that various substances, such as polymers [1], flavonoids [2] and gallic acid [3], upon heating, do not exhibit the typical thermophysical transition of melting, and instead, they decompose. In many cases, the decomposition and "melting" (material softening or fluidization) overlap and occur simultaneously. The term "thermochemical transition" was proposed for this effect [1]. In some cases, e.g., gallic acid [3], in addition to the solid–liquid thermochemical transition at temperatures above 200 °C, a solid–solid thermochemical transition was observed to occur at temperatures below 100 °C. This behavior was explained based on the comparison of the Gibbs free energies of decomposition and melting and was put in the broader frame of "melting inability" [4].

In addition to the term melting inability, the term "desolvation inability" was introduced in order to describe the occurrence of decomposition of a solute prior to being able to accomplish the solvent's full removal upon the heating of a solvate/hydrate [5].

An explanation, similar to that for melting inability, has been provided for the desolvation inability; that is, the Gibbs free energy of decomposition is lower than the one of desolvation/dehydration [5]. The desolvation inability is the actual cause of the above-mentioned solid–solid thermochemical transition of gallic acid below 100 °C. In substances like flavonoids and gallic acid, which exhibit antioxidant properties, it would be interesting to examine if and how the antioxidant activity is influenced by the minor decomposition that occurs during the solid–solid thermochemical transition. For substances with melting and/or desolvation inability, a confusion regarding their thermal behavior has been recognized. For example, the solid–solid thermochemical transition of gallic acid has been interpreted as water evaporation or solid–solid crystallographic (thermophysical) transformation [3]. Also, some confusion was reported for its infrared (IR) absorbance bands [3]. Of course, the knowledge of the IR bands for a given a substance is very useful on many occasions.

The primary decomposition pathway for gallic acid was found to be an esterification reaction. Also, it has been reported that the thermal behavior of cellulose esters is governed to great extent by the esterification reaction between the polymer and its acid impurities, which can be produced by the hydrolysis of the polymer [1]. The easiness of the occurrence of the esterification reaction was reported to be related to the low value of the enthalpy of esterification [1]. As mentioned above, a thermodynamic explanation has been reported for the melting inability, and also this behavior has been predicted for various organic substances including various hydroxyl-acids, e.g., citric acid [4]. Citric acid is a naturally occurring hydroxy-acid that is widely used as an antioxidant, a preservative, a stabilizer, etc., in the Food [6,7] and Pharmaceutical [8,9] industry. Citric acid is believed to exist in anhydrous and monohydrate forms. Confusion regarding the thermal behavior of citric acid can be recognized in the literature. Although it has been reported that the melting and decomposition of anhydrous citric acid occur at very close, practically overlapping, temperature ranges [10], in various cases [11–14], including the NIST Chemistry Webbook [12] and the Handbook of Chemistry and Physics [11], typical melting points are reported in the range of 150–160 °C. In the Handbook of Chemistry and Physics, a melting point of 135 °C is reported for citric acid monohydrate and a melting point of 153 °C is reported for anhydrous citric acid [11]. Also, no information is given for the boiling point of the monohydrate form, while for the anhydrous form, is it mentioned that it decomposes and does not boil [11]. The monohydrate form, in addition to the melting peak, exhibits an endothermic peak in its Differential Scanning Calorimetry (DSC) curve below 100 °C [15,16]. Anhydrous citric acid is believed to be able to be prepared by heating, e.g., drying at 100 °C to remove absorbed moisture [10], while citric acid monohydrate is believed to dehydrate, and it has been reported that the onset temperature of its dehydration is 71 °C [16]. Of course, if the monohydrate form is able to dehydrate, then it should not exhibit any melting point, and there should not be any difference in the melting point of the anhydrous and hydrate form. However, as mentioned above, in the Handbook of Chemistry and Physics quite different values are given for the melting points.

In addition to the thermal behavior, an interesting aspect can be recognized in the IR spectrum of citric acid and citric acid monohydrate. More precisely, in the C=O stretching region (~1650–1800 cm^{-1}), citric acid monohydrate exhibits a triple peak [17], while citric acid exhibits a double peak [12]. However, it is widely known [18] that the C=O stretching IR bands are very characteristic depending on the origin of the C=O group (that is, the C=O stretching of ketones, acids, esters, etc., appears at distinct and characteristic wavenumbers). In both citric acid and citric acid monohydrate, only one type (acid) of the C=O group is supposed to exist; thus, only one respective peak should be present in the IR spectrum. However, as mentioned above, this is not the case. In addition, in the IR spectrum of citric acid, one sharp peak in the O-H stretching region (~3100–3600 cm^{-1}) is present [12]. Such sharp peaks are not typical of the O-H bands of organic molecules. In addition, this sharp peak seems to be the same peak that was observed in gallic acid and was assigned to water O-H stretching. In a few words, there are various aspects that require clarification regarding the thermal behavior and the IR absorbance bands of citric acid and its monohydrate form. The scope of this work is to provide insights regarding these issues. More precisely, we provide a comprehensive analysis of the thermal behavior and the infrared spectrum of citric acid monohydrate and citric acid. We found that citric acid monohydrate cannot be dehydrated without being decomposed, and neither the hydrate nor the non-hydrate form exhibits a melting point. It is shown that citric acid does not exists in pure form but only as a mixture of citric acid, water and a citric acid oligomer that is produced through esterification. These strongly affect its IR spectrum. It is also shown that the partial alteration of the chemical structure of citric acid can occur easily by mild thermal treatment or even at room temperature. The alteration of the chemical structure affects its antioxidant activity.

2. Materials and Methods

Citric acid monohydrate (Food Grade) was purchased from a local market (Volos, Greece). KBr (>99.5%) was purchased from Chem-Lab (Zedelgem, Belgium). D_2O (99.9%) was purchased from Deutero GmbH (Kastellaun, Germany). Trolox (6-hydroxy-2,5,7,8-tetramethylchroman-2-carboxylic acid) was purchased from Sigma (St. Louis, MO, USA). The KBr was dried at 140 °C for 2 h prior to use, while all other chemicals were used as received.

Thermogravimetric analysis (TGA) measurements were carried out using a Shimadzu TGA-50 thermogravimetric analyzer (Shimadzu, Tokyo, Japan). Differential Scanning Calorimetry (DSC) measurements were performed using a Shimadzu DSC-50 calorimeter (Shimadzu, Tokyo, Japan). Fourier Transform Infrared (FTIR) measurements were performed using a Thermo FTIR spectrometer (Thermo Fisher Scientific, Waltham, MA, USA). Antioxidant activity was evaluated spectrophotometrically. A Christ freeze dryer (model Gamma 1-20, Martin Christ Gefriertrocknungsanlagen, Osterode am Harz, Germany) was used for the preparation of the freeze-dried sample.

Four samples were examined, namely, (a) raw citric acid monohydrate, (b) citric acid monohydrate heated at 100 °C, (c) citric acid monohydrate freeze-dried and (d) citric acid recrystallized from a D_2O solution. The recrystallized sample was prepared by solution casting. The solution was left at room temperature for 3 days for complete solvent evaporation. For the freeze-drying process, initially the sample was frozen at −20 °C under atmospheric pressure. Then, vacuum was applied, and simultaneously the plate in which the sample was placed was heated in order to provide the latent heat of sublimation. During freeze drying, the temperature of the drier was stabilized at −12 °C.

The raw citric acid sample was examined with DSC, TGA and FTIR. All other samples were examined with FTIR. Aqueous solutions of the raw and the heated samples were examined for their antioxidant activity. The Ferric Reducing/Antioxidant Power (FRAP) assay was performed using the modified method by Pulido et al. [19]. In brief, 100 μL of the raw and heated samples (0.1 g/mL) was mixed with 2.9 mL of freshly prepared FRAP reagent. Samples were mixed and incubated at 37 °C for 10 min, and the absorbance of each sample was measured at 593 nm. Trolox was used as a standard for the calibration curve. The results were expressed in μM of Trolox equivalents. The DSC and TGA measurements were performed with a heating rate of 10 K/min from room temperature up to 165 °C under nitrogen atmosphere (20 mL/min flow). The FTIR measurements were carried out with a resolution of 2 cm^{-1} by performing 64 scans in absorbance mode. Baseline correction and normalization at a scale of 0–1 was applied in all the FTIR spectra. In order to facilitate the comparison of the spectra of the samples, the approach of spectra subtraction was adopted.

3. Ab Initio DFT Calculations

One of the IUPAC names of citric acid is 3-carboxy-3-hydroxypentane-1,5-dioic acid [12]. We used this name as the basis in order to give code names to the possible decomposition products of citric acid. According to this name, it can be realized that citric acid has three COOH groups at positions 1, 3 and 5 and one OH group at position 3. Since there are three

COOH groups and one OH group in citric acid, there are two possible different esters that can be produced (by considering the COOH groups at positions 1 and 5 equivalent). In addition, the esterification reaction can take place either inter-molecularly (that is, between two different citric acid molecules) to give a dimer or intra-molecularly (that is, within the same molecule) to give a cyclic substance. Also, a dehydration reaction between the COOH groups is possible (again both inter- and intra-molecularly) to give anhydride. In total, there are nine possible reactions with nine decomposition products along with water as the second decomposition product. In order to avoid complex and large names and in order to facilitate the presentation and discussion of the results, we give a code name to these products. The code name is composed of one letter, two numbers and the term "dim", e.g., E13dim, A15. The letter may be A or E, corresponding to an anhydride or ester product, respectively. In the anhydrides, the numbers correspond to the positions of the COOH groups that take part in the reaction. In esters, the first number corresponds to the position of the COOH group, while the second number is 3, corresponding to the position of the OH group. The term "dim" refers to dimers and thus is used for the products produced by inter-molecular reactions. For example, the name A13dim corresponds to the anhydride that is formed through the reaction of the COOH group at position 1 of one citric acid molecule with the COOH group at position 3 of another molecule. The name E13 is the cyclic ester produced by the reaction of the COOH group at position 1 with the OH group at position 3 within the same citric acid molecule. As mentioned above, the COOH at position 1 and 5 are considered equivalent, and thus, the E53dim is the same as the E13dim.

In this work, we used DFT to find the Gibbs free energy of the above-mentioned nine reactions in order to determine which of the nine products is the most probable product. According to the literature, the PBE0 functional with a TZ basis size is accurate for structure and frequency optimization and of intermediate accuracy, as is the other functional for estimating energies of reactions [20]. Also, it is recommended to use a dispersion correction in the calculations [20]. Thus, we used the PBE0 [21,22] functional with the def2-TZVP [23] basis set along with the def2/J [24] auxiliary basis and the D4 [25,26] dispersion correction.

The molecules were drawn in Avogadro software (Windows version 1.1.0) [27,28], and an initial (rough) geometry optimization was also performed with Avogadro software. These atom coordinates were used as input for the geometry optimization with ORCA software [29–31]. The frequency optimization was also performed with ORCA software [29–31]. For the calculations, ORCA uses the libint2 library for the computation of 2-el integrals [32]. Rotational and vibrational entropy was computed in ORCA according to Herzberg [33] and the QRRHO of Grimme [34], respectively. We used version 5.0.2 of ORCA, which was built with support for libXC version 5.1.0 [35].

4. Results and Discussion

4.1. *Raw and Heated at 100 °C Citric Acid Monohydrate Samples*

4.1.1. Thermal Behavior and IR Spectra

In Figure 1, the DSC and TGA curves of citric acid monohydrate are presented in the range of 40–165 °C. As can be seen, in this temperature range, citric acid monohydrate exhibits two endothermic effects (DSC curve), and both of them overlap with considerable mass loss (TGA curve). The first thermal effect is centered at 61 °C, while the second effect is located at 150 °C.

According to the established interpretations, we could explain the above results as follows: citric acid monohydrate dehydrates, the mass loss up to 90 °C is water evaporation, and then the anhydrous citric acid melts at 150 °C [11,12]. However, we dispute that such an interpretation is correct. Although the temperature of 150 °C is very close to the literature values [11,12] of the melting point of citric acid, it is clear that this effect overlaps with mass loss. In Figure 2, photographs are presented that show inside the TGA pan of the raw citric acid monohydrate sample and the same sample after being heated up to 165 °C (the temperature of 165 °C was selected since at this temperature the thermal effect is completed, as can be seen in the DSC curve in Figure 1).

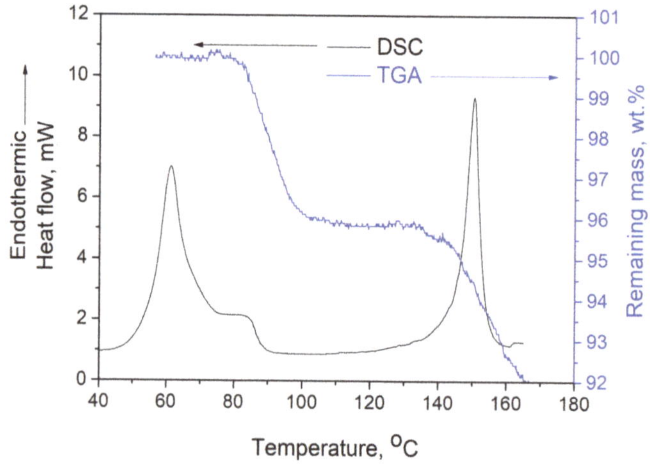

Figure 1. DSC and TGA curves of raw citric acid monohydrate.

(a) (b)

Figure 2. Photographs of (**a**) raw citric acid sample and (**b**) the citric acid monohydrate sample heated at 165 °C.

It is clear from the photos of Figure 2 that the sample undergoes severe decomposition by heating up to 165 °C. As mentioned above, the center of the second peak in the DSC curve is located at 150 °C, and the end of this thermal effect is located in the range of 160–165 °C. If this thermal effect was melting, then no sign of decomposition should be observed after heating the sample at this temperature range. Thus, it is clear that the thermal effect that is centered at 150 °C and ends at 165 °C cannot be melting. Citric acid exhibits melting inability. Actually, it is not citric acid that exhibits melting inability but a mixture of citric acid, water and the ester of citric acid, as we will discuss next. More precisely, we will now argue that the broad endothermic peak at 61 °C does not arise solely from the heat, which is required for water evaporation, but there is a contribution from the heat, which is required for the esterification reaction that occurs upon the heating of the sample. In order to support this claim, we will focus on the thermal effect below 100 °C, which is centered at 61 °C. In order to study this effect, a citric acid monohydrate sample was heated up to 100 °C. The temperature of 100 °C was selected since at this temperature, the mass loss of the first stage has been completed (see TGA curve in Figure 1) and the thermal effect at 61 °C seems to be completed (see DSC curve in Figure 1). The FTIR spectra of the raw and the heated at 100 °C samples along with their subtraction spectrum are presented in Figure 3.

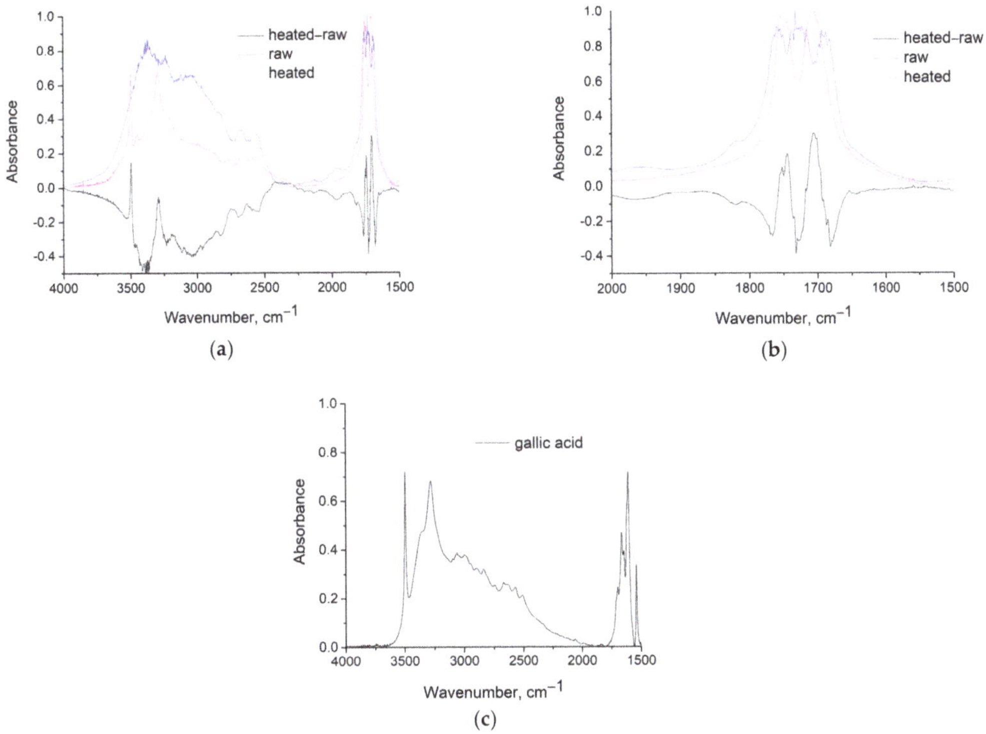

Figure 3. (**a**) FTIR spectra of the raw and the heated at 100 °C samples and their subtracted spectra (heated-raw) in the 1500–4000 cm^{-1} region, (**b**) Same spectra as in (**a**) in the 1500–2000 cm^{-1} region and (**c**) FTIR spectra of gallic acid.

As can be seen in Figure 3a, the IR spectrum of the two samples is quite different. Obvious alterations can be seen in the region of O-H stretching (3100–3700 cm^{-1}). Also, in Figure 3b, the same spectra are presented around the region of C=O stretching in order to show the double peak of the heated sample and the triple peak of the monohydrate sample, which were mentioned in the Introduction. The presence of multiple distinct C=O stretching vibrations indicates the existence of multiple types of C=O. The wavenumbers of the triple peak are 1690, 1727 and 1753 cm^{-1}. The peak at the higher wavenumber at 1753 cm^{-1} cannot be assigned to acid groups since it is characteristic of aliphatic ester C=O or anhydride C=O, which typically appear at even higher wavenumbers, e.g., 1740–1800 cm^{-1} [18]. It is known that the C=O stretching vibration of carboxylic acids appears at quite lower wavenumbers, e.g., 1700 cm^{-1} [18]. Thus, the peak at 1753 cm^{-1} cannot be assigned to acid C=O and points out the existence of ester C=O. In addition, the water bending vibration occurs at around 1650 cm^{-1}, but no such peak is present in the spectrum of citric acid monohydrate. In the case of gallic acid, shifting of the water bending peak was observed, which depended on if the water molecule interacted with the COOH or OH group [3]. Thus, it seems likely that the water bending band in the spectrum of the citric acid monohydrate is shifted at higher wavenumbers and overlaps with the C=O stretching of acid groups. Thus, the triple peak of the citric acid monohydrate in the C=O region contains contributions from acid C=O stretching, ester C=O stretching and water H-O-H bending. This means that the citric acid monohydrate does not exist in the solid state, it partially decomposes through esterification, and the substance we name citric acid monohydrate is actually a mixture of citric acid, water and citric acid ester, most likely in equilibrium. In the region of O-H stretching (3100–3700 cm^{-1}), the characteristic bands of water O-H stretching at 3490 and 3280 cm^{-1} [18] cannot be distinguished in the IR spectrum of citric acid monohydrate since they overlap with the O-H stretching of the OH group at higher wavenumbers and the O-H stretching of COOH groups, which is known to occur at lower wavenumbers, e.g., in the region 2500–3300 cm^{-1} [18].

In the heated sample, two sharp peaks at 3494 and 3290 cm^{-1} are present. The two peaks are very close to the literature values of water at 3490 and 3280 cm^{-1} [18]. These two peaks were also present in the IR spectrum of gallic acid and were found to be related to water O-H stretching [3]. The IR spectrum of gallic acid is presented in Figure 3c, in which the resemblance with the spectrum of the heated citric acid can be seen in the region 3000–3600 cm^{-1}. The above-mentioned peaks and especially the sharp peak at 3498 cm^{-1} are clearly visible in both spectra. In the subtracted spectrum (Figure 3a,b), it is clear that these are two positive peaks inside the overall negative O-H stretching peak. This clearly points out that the overall OH groups have decreased, with a simultaneous increase in the water content. The esterification reaction could explain this alteration in the spectrum of the heated sample, since it should result in a decrease in O-H stretching of the COOH and OH groups and cause an increase in the water bands. Of course, in such a case, the water bending band should also be positive, and the ester C=O band should also increase. Indeed, in the subtracted spectrum, it can be seen that two positive peaks exist at 1745 cm^{-1} (assigned to ester C=O) and at 1705 cm^{-1} (assigned to water H-O-H bending). The negative peak at this region can be related to the reduction of the acid C=O groups due to the esterification reaction. Based on the above, we propose the band assignments that are presented in Tables 1 and 2 for citric acid monohydrate and citric acid anhydrous (heated at 100 °C), respectively.

Table 1. Proposed IR band assignments for citric acid monohydrate in the region 1650–4000 cm^{-1}.

Wavenumber, cm^{-1}	Citric Acid Monohydrate
3340–3700	Overlapping of O-H stretching vibration of water and alcohol OH groups; the contribution of water is around 3490 cm^{-1}
3000–3340	O-H stretching vibration of acid OH groups with water contribution at around 3280 cm^{-1}
1745–1800	C=O stretching of ester carbonyl
1704–1745	C=O stretching of acid carbonyl overlapping with ester C=O at higher wavenumbers
1650–1704	Overlapping of H-O-H bending with C=O stretching of acid carbonyl

Table 2. Proposed IR band assignments for citric acid anhydrous (citric acid monohydrate heated at 100 °C) in the region 1650–4000 cm^{-1}.

Wavenumber, cm^{-1}	Citric Acid
3498	O-H stretching vibration of water
3445	O-H stretching vibration of alcohol OH
3291	Mostly O-H stretching vibration of water overlapping with acid OH groups
3000–3340	O-H stretching vibration of acid OH groups
1727–1800	C=O stretching of ester carbonyl
1650–1727	Overlapping of H-O-H bending with C=O stretching of acid carbonyl

As a summary of this section, we can say that citric acid monohydrate undergoes esterification to some extent even at room temperature during its preparation/storage. Upon heating, it exhibits desolvation inability and does not dehydrate. Instead, by heating, the equilibrium between citric acid, its ester and water is disturbed and is shifted towards the direction of esterification, leading to the formation of more citric acid ester. This mixture ("anhydrous citric acid") exhibits melting inability, and any fluidization occurs simultaneously with decomposition upon further heating.

4.1.2. Antioxidant Activity

As discussed in the previous section, the citric acid monohydrate sample undergoes a structure alteration through esterification at temperatures below 100 °C. Such temperatures may be used during food or drug processing where citric acid is used as an antioxidant, flavoring, a preserving agent, etc. Thus, we examined the antioxidant activity of aqueous solutions of the raw and the heated at 100 °C citric acid samples. The antioxidant capacity of the raw sample was found to be equal to 257 μM TE, while the heated sample exhibited a higher capacity of 312 μM TE. It should be stressed that citric acid does not have a direct antioxidant activity [36]; that is, it is not a primary antioxidant, but it can act as secondary antioxidant by inhibiting oxidation through chelation [36,37]. The antioxidant activity/chelation in this assay (which involves iron reduction) is related to the increased number of OH groups present [37], and since esterification leads to the reduction of such groups, the opposite result would be expected. The increase in the antioxidant capacity of the heated sample may be understood in terms of the increase in the molecular weight. More precisely, the esterification may take place between various citric acid molecules, and an oligomer (e.g., dimer, trimer, etc.) may be formed. Per mole of citric acid monohydrate, theoretically there are 4 moles of OH groups. Per mole of citric acid dimer, there are 6 moles of OH groups. This increase may explain the increase in the antioxidant activity. Also, it must be taken into account that inside water, the excess of water, according to Le Chatelier's Principle, may favor the reverse reaction (hydrolysis), and more than six OH groups may be presented per mole of the dimer after hydrolysis. Finally, the different water content of the samples may also affect the observed alteration.

4.2. Recrystallized Sample from D_2O Solution

In Figure 4a, the IR spectra of the raw sample and the recrystallized citric acid sample from the D_2O solution are presented, along with their subtracted spectrum. No signs of the presence of D_2O can be detected (that is, the D-O stretching at 2540 and 2450 cm^{-1} and the D-O-D bending at 1215 cm^{-1} [18]). Interestingly, the spectrum of the recrystallized sample is pretty similar to the spectrum of the heated at 100 °C sample and similar conclusions are derived from the comparison of the spectra of the recrystallized and the raw sample. That is, during solvent evaporation at room temperature, it seems that esterification took place, leading to alterations in the spectrum similar to the ones caused by heating at 100 °C. In an excess of water, as mentioned above, the favoring of the hydrolysis direction is expected. However, as the evaporation of water proceeds and its concentration is decreased, the ongoing evaporation of water, according to Le Chatelier's Principle, shifts the reaction towards the esterification direction. This could explain the coincidence of the spectra of the heated and recrystallized samples.

As mentioned in the previous section, if the esterification proceeds between different citric acid molecules (and not within the same molecule to give a cyclic substance), then an increase in the molecular weight is expected due to the formation of oligomers. Indeed, after recrystallization, as shown in Figure 4b, the sample was not in a powder form (behavior typical of low-molecular-weight substances) but in a film-like form (typical behavior of polymers). Of course, the film did not exhibit the mechanical coherence of typical macromolecules, but this poor ability for film formation clearly points out an increase in the molecular weight. This can be considered as additional evidence for the occurrence of inter-molecular esterification.

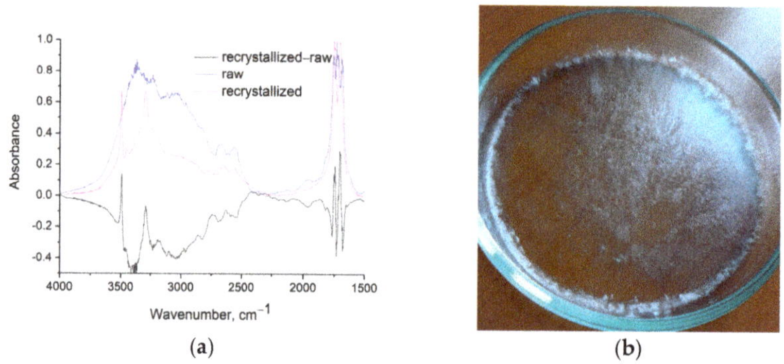

Figure 4. (**a**) FTIR spectra of the raw and the recrystallized samples and their subtracted spectra (recrystallized-raw) and (**b**) photograph of the recrystallized sample after solvent evaporation showing a film poor formation.

4.3. Freeze-Dried Sample

In Figure 5a, the IR spectra of the raw and the freeze-dried citric acid monohydrate sample are presented, along with their subtracted spectrum. As can be seen, the differences in the two spectra are less severe compared to the ones of the heated and the recrystallized samples. Since we have concluded that heating favors the esterification reaction, it would be expected that the decrease in temperature will favor the reverse reaction, and thus it is expected that esterification would proceed at a lower extent in the case of freeze drying. In addition, due to the low temperatures that are used during freeze drying, the reaction rate is expected to be quite lower. From the subtracted spectra, it is clear that only one positive peak exists at 1726 cm^{-1}. This peak could be attributed to a very small increase in the water bending band or/and the C=O ester stretching band. In such case, this means that a very minor decomposition has occurred during freeze drying. It is worth mentioning that in the case of gallic acid, clear signs of decomposition during freeze drying have been reported [5]. In any case, there is no evidence in the FTIR spectra to claim that water has been removed or even decreased after 21 h of freeze drying. This is another indication of the desolvation inability of citric acid monohydrate. For comparative analysis, in Figure 5b, the spectra of the raw, recrystallized and freeze-dried samples are presented.

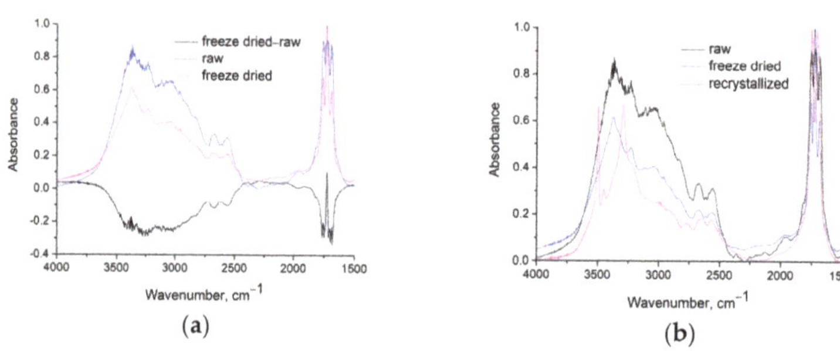

Figure 5. (**a**) FTIR spectra of the raw and the freeze-dried samples and their subtracted spectra (freeze dried-raw) and (**b**) FTIR of the raw, recrystallized and freeze-dried samples.

4.4. DFT Calculations

In Table 3, the values of the Gibbs free energy, as calculated from DFT, of the nine possible reactions of citric acid esterification or anhydride formation are presented. As can be seen, seven out of the nine reactions have ΔG values in the order of magnitude of 10–100 kJ/mol, one reaction has a ΔG more than 100 kJ/mol, and one reaction has a very low (as an absolute value) ΔG value close to 1 kJ/mol. This reaction is the reaction that leads to the formation of a dimer through esterification between the OH group at position 3 of one molecule and the COOH group at position 3 of another citric acid molecule. Thus, the most probable product is the E33dim. This is in agreement with the interpretations of the experimental results for the formation of oligomers and the increase in molecular weight. Also, despite the fact that the ΔG values, as calculated from DFT, refer to gas-phase reactions, the very low value of ΔG implies and may explain the easiness of the occurrence of this reaction. This is also in agreement with the conclusions of the previous sections, where it was claimed that the esterification reaction takes place even at room temperature. In addition to the E33dim product, other probable products are mainly cyclic anhydrides, e.g., A13 and A15. These products may be formed during the decomposition at 150 °C or to a lower extent at lower temperatures.

Table 3. ΔG of the nine possible decomposition reactions of citric acid (formation of ester or anhydride) as calculated from DFT.

Product	ΔG of Reaction, kJ/mol
E33dim	+1.6
A13	+13.9
A15	+29.1
A13dim	+42.9
E13	+44.5
A11dim	+46.1
A33dim	+57.3
E13dim	+58.5
E33	+115.2

5. Further Discussion

The results of this study showed that citric acid undergoes alterations of its chemical structure rather easily by a mild thermal treatment or even at room temperature. Since the esterification reaction is a two-direction reaction and can be reversed, the extent of esterification depends on temperature, available water, air humidity, etc. Thus, citric acid can undergo different alterations of its structure depending on its storage conditions, subjected treatment, etc. The provided insights could assist in improving the performance and quality of food or drug formulations that involve the use of citric acid. More precisely, based on the results of this study, it is clear that citric acid does not always exhibit the very same properties, which seem to depend on the subjected treatment as well as its storage conditions. For achieving the optimum design of a food or drug formulation, it is necessary to use the proper amounts of additives such as preservatives, antioxidants, etc.; that is, there should not be an amount of additive higher than the one needed, and on the other hand, the amount of the additive should be sufficient in order to be active and adequate to fulfill its purpose of use. The alteration of the chemical structure of citric acid obviously affects its properties either directly or indirectly. An example of direct influence, as shown in this study, is the alteration of its antioxidant activity, which can be related to the ability for chelation capacity. The alteration of the chemical structure due to various treatment or storage conditions affects the pH of citric acid and thus can also affect, indirectly, its preservative activity, which is related to its pH value. Similarly, citric acid is used to improve the solubility/release of drugs that are poorly soluble in water. The alteration of the structure and the reduction of OH and COOH groups can also affect the solubility and hydrophilicity of citric acid. All in all, different batches of citric acid may exhibit different chemical structures and thus different properties. Thus, such issues should be taken into account in order to achieve the optimum performance/quality of a product.

6. Conclusions

Citric acid does not seem to exist in pure form in the solid state either in the hydrate or anhydrous form. In the solid state, it exists as a mixture. More precisely, the multiple bands of the IR spectrum in the C=O stretching region point outs that what we call citric acid monohydrate is a mixture of citric acid, the citric acid oligomer produced through esterification and water. This complex hydrate exhibits desolvation inability, and upon heating it decomposes through esterification, leading to the formation of more ester product. Citric acid subjected to heating contains water and ester product. It exhibits melting inability (does not exhibit any melting point), and upon heating any fluidization occurs simultaneously with decomposition. In addition to the experimental evidence, ab initio DFT calculations indicate that the most probable product is the one produced through esterification between the OH group at position 3 of one molecule and the COOH group at position 3 of another citric acid molecule (the positions refer to the following IUPAC name of citric acid: 3-carboxy-3-hydroxypentane-1,5-dioic acid). Other probable decomposition products are cyclic anhydrides. The increase in the molecular weight during esterification along with the reversion of the reaction inside water may explain the increase in the antioxidant activity of the heated ("anhydrous") sample over the one of the raw citric acid monohydrate sample. The monohydrate form cannot be dehydrated either by heating or by freeze drying.

In few words, citric acid undergoes alterations of its chemical structure by mild heating or even at room temperature. The primary reaction that is responsible for this alteration is an esterification reaction. The extent of esterification (which is a two-direction reaction and can be reversed to some extent) depends on temperature or an excess of products or reactants and thus depends on the subjected treatment and storage conditions. Thus, citric acid samples that have been subjected to different treatment or storage conditions are likely to exhibit different structures and thus different properties, e.g., different antioxidant activities. The provided insights could be useful for optimizing the design and performance of food and drug formulations that involve the use of citric acid.

Author Contributions: Conceptualization, C.T.; methodology, C.T. and P.M.; validation, C.T., A.P. and P.M.; formal analysis, C.T., A.P. and P.M.; investigation, C.T., A.P. and P.M.; resources, C.T. and P.M.; data curation, C.T.; writing—original draft preparation, C.T.; writing—review and editing, C.T. and P.M.; visualization, C.T.; supervision, C.T. and P.M. All authors have read and agreed to the published version of the manuscript.

Funding: This research received no external funding.

Institutional Review Board Statement: Not applicable.

Informed Consent Statement: Not applicable.

Data Availability Statement: Data is contained within the article.

Acknowledgments: The authors would like to thank Ioannis Tsivintzelis for providing chemicals and access to the instruments of the Laboratory of Physical Chemistry of the Chemical Engineering Department of the Aristotle University of Thessaloniki, E. Tzimpilis for his assistance with the preparation of the recrystallized sample, S. Matsia for her assistance with the FTIR measurements and A. Goulas for his assistance with the freeze-drying process.

Conflicts of Interest: The authors declare no conflicts of interest.

References

1. Tsioptsias, C.; Nikolaidou, E.G.; Ntampou, X.; Tsivintzelis, I.; Panayiotou, C. Thermo-chemical transition in cellulose esters and other polymers. *Thermochim. Acta* **2022**, *707*, 179106. [CrossRef]
2. Tsioptsias, C.; Spartali, C.; Marras, S.I.; Ntampou, X.; Tsivintzelis, I.; Panayiotou, C. Thermochemical Transition in Low Molecular Weight Substances: The Example of the Silybin Flavonoid. *Molecules* **2022**, *27*, 6345. [CrossRef]
3. Tsioptsias, C.; Tsivintzelis, I. Insights on thermodynamic thermal properties and infrared spectroscopic band assignments of gallic acid. *J. Pharm. Biomed. Anal.* **2022**, *221*, 115065. [CrossRef]

4. Tsioptsias, C. Thermodynamic explanation and criterion for the exhibition of melting inability in molecular species. *AIMS Mater. Sci.* **2023**, *10*, 618–636. [CrossRef]
5. Tsioptsias, C. Desolvation Inability of Solid Hydrates, an Alternative Expression for the Gibbs Free Energy of Solvation, and the Myth of Freeze-Drying. *Materials* **2024**, *17*, 2508. [CrossRef]
6. Rodríguez-Aguilar, F.; Ortega-Regules, A.E.; Ramírez-Rodrigues, M.M. Influence of time-temperature in the antioxidant activity, anthocyanin and polyphenols profile, and color of Ardisia compressa K. extracts, with the addition of sucrose or citric acid. *Food Chem.* **2024**, *440*, 138181. [CrossRef] [PubMed]
7. Zhu, H.; Gao, L.; Liang, J.; Erihemu; Li, G.; Song, X.; Qi, W. Applications and characterization of anti-browning enzymatically modified potato starch (EPS) film associated with chitosan (CTS)/L-Cys/citric acid (CA) on fresh-cut potato slices. *Food Chem.* **2024**, *452*, 139424. [CrossRef]
8. Lu, Q.; Dun, J.; Chen, J.-M.; Liu, S.; Sun, C.C. Improving solid-state properties of berberine chloride through forming a salt cocrystal with citric acid. *Int. J. Pharm.* **2019**, *554*, 14–20. [CrossRef]
9. Onoue, S.; Yamamoto, K.; Kawabata, Y.; Hirose, M.; Mizumoto, T.; Yamada, S. Novel dry powder inhaler formulation of glucagon with addition of citric acid for enhanced pulmonary delivery. *Int. J. Pharm.* **2009**, *382*, 144–150. [CrossRef]
10. Barbooti, M.M.; Al-Sammerrai, D.A. Thermal decomposition of citric acid. *Thermochim. Acta* **1986**, *98*, 119–126. [CrossRef]
11. Haynes, W.M. (Ed.) *Handbook of Chemistry and Physics*, 97th ed.; CRC Press LLC: Boca Raton, FL, USA, 2016.
12. NIST. Available online: https://webbook.nist.gov/cgi/cbook.cgi?ID=77-92-9 (accessed on 15 May 2024).
13. Willfahrt, A.; Steiner, E.; Hötzel, J.; Crispin, X. Printable acid-modified corn starch as non-toxic, disposable hydrogel-polymer electrolyte in supercapacitors. *Appl. Phys. A* **2019**, *125*, 474. [CrossRef]
14. Wyrzykowski, D.; Hebanowska, E.; Nowak-Wiczk, G.; Makowski, M.; Chmurzyński, L. Thermal behaviour of citric acid and isomeric aconitic acids. *J. Therm. Anal. Calorim.* **2011**, *104*, 731–735. [CrossRef]
15. Zhao, S.; Feng, F.; Yu, F.; Shen, Q. Flower-to-petal structural conversion and enhanced interfacial storage capability of hydrothermally crystallized MnCO3via the in situ mixing of graphene oxide. *J. Mater. Chem. A* **2015**, *3*, 24095–24102. [CrossRef]
16. Groen, H.; Roberts, K.J. Nucleation, Growth, and Pseudo-Polymorphic Behavior of Citric Acid As Monitored in Situ by Attenuated Total Reflection Fourier Transform Infrared Spectroscopy. *J. Phys. Chem. B* **2001**, *105*, 10723–10730. [CrossRef]
17. Li, S.; Cai, G.; Wu, S.; Raut, A.; Borges, W.; Sharma, P.R.; Sharma, S.K.; Hsiao, B.S.; Rafailovich, M. Sustainable Plant-Based Biopolymer Membranes for PEM Fuel Cells. *Int. J. Mol. Sci.* **2022**, *23*, 15245. [CrossRef]
18. Stuart, B. *Infrared Spectroscopy: Fundamentals and Applications*; John Wiley and Sons Ltd.: West Sussex, UK, 2004.
19. Pulido, R.; Bravo, L.; Saura-Calixto, F. Antioxidant Activity of Dietary Polyphenols As Determined by a Modified Ferric Reducing/Antioxidant Power Assay. *J. Agric. Food Chem.* **2000**, *48*, 3396–3402. [CrossRef]
20. Bursch, M.; Mewes, J.-M.; Hansen, A.; Grimme, S. Best-Practice DFT Protocols for Basic Molecular Computational Chemistry. *Angew. Chem. Int. Ed.* **2022**, *61*, e202205735. [CrossRef]
21. Perdew, J.P.; Ernzerhof, M.; Burke, K. Rationale for mixing exact exchange with density functional approximations. *J. Chem. Phys.* **1996**, *105*, 9982–9985. [CrossRef]
22. Adamo, C.; Barone, V. Toward reliable density functional methods without adjustable parameters: The PBE0 model. *J. Chem. Phys.* **1999**, *110*, 6158–6170. [CrossRef]
23. Weigend, F.; Ahlrichs, R. Balanced basis sets of split valence, triple zeta valence and quadruple zeta valence quality for H to Rn: Design and assessment of accuracy. *Phys. Chem. Chem. Phys.* **2005**, *7*, 3297–3305. [CrossRef]
24. Weigend, F. Accurate Coulomb-fitting basis sets for H to Rn. *Phys. Chem. Chem. Phys.* **2006**, *8*, 1057–1065. [CrossRef] [PubMed]
25. Caldeweyher, E.; Bannwarth, C.; Grimme, S. Extension of the D3 dispersion coefficient model. *J. Chem. Phys.* **2017**, *147*, 034112. [CrossRef] [PubMed]
26. Caldeweyher, E.; Ehlert, S.; Hansen, A.; Neugebauer, H.; Spicher, S.; Bannwarth, C.; Grimme, S. A generally applicable atomic-charge dependent London dispersion correction. *J. Chem. Phys.* **2019**, *150*, 154122. [CrossRef] [PubMed]
27. Software Avogadro. *An Open-Source Molecular Builder and Visualization Tool*, Version 1.1.0; Avogadro: Fontenilles, France, 2009. Available online: http://avogadro.cc/.
28. Hanwell, M.D.; Curtis, D.E.; Lonie, D.C.; Vandermeersch, T.; Zurek, E.; Hutchison, G.R. Avogadro: An advanced semantic chemical editor, visualization, and analysis platform. *J. Cheminform.* **2012**, *4*, 17. [CrossRef]
29. Neese, F. The ORCA program system. *WIREs Comput. Mol. Sci.* **2012**, *2*, 73–78. [CrossRef]
30. Neese, F. Software update: The ORCA program system, version 4.0. *WIREs Comput. Mol. Sci.* **2018**, *8*, e1327. [CrossRef]
31. Neese, F.; Wennmohs, F.; Becker, U.; Riplinger, C. The ORCA quantum chemistry program package. *J. Chem. Phys.* **2020**, *152*, 224108. [CrossRef]
32. Libint. Available online: http://libint.valeyev.net.
33. Herzberg, G. *Molecular Spectra and Molecular Structure. Part II. Infrared and Raman Spectra of Polyatomic Molecules*; Van Nostrand Comp.: New York, NY, USA, 1945.
34. Grimme, S. Supramolecular Binding Thermodynamics by Dispersion-Corrected Density Functional Theory. *Chem.—A Eur. J.* **2012**, *18*, 9955–9964. [CrossRef]
35. libXC. Available online: https://tddft.org/programs/libxc/ (accessed on 15 May 2024).

36. Chahardoli, A.; Jalilian, F.; Memariani, Z.; Farzaei, M.H.; Shokoohinia, Y. Chapter 26—Analysis of organic acids. In *Recent Advances in Natural Products Analysis*; Sanches Silva, A., Nabavi, S.F., Saeedi, M., Nabavi, S.M., Eds.; Elsevier: Amsterdam, The Netherlands, 2020; pp. 767–823. [CrossRef]
37. Amarowicz, R.; Pegg, R.B. Chapter One—Natural antioxidants of plant origin. In *Advances in Food and Nutrition Research*; Ferreira, I.C.F.R., Barros, L., Eds.; Academic Press: Cambridge, MA, USA, 2019; Volume 90, pp. 1–81.

applied sciences

MDPI

Article

Convection and Microwave–Convection Drying of Moldavian Dragonhead (*Dracocephalum moldavica* L.) Leaves

Ewa Gondek [1,*], Anna Kamińska-Dwórznicka [1], Sławomir Kocira [2,3], Tomasz Oniszczuk [4], Michał Bialik [1] and Mateusz Stasiak [5]

1 Institute of Food Sciences, Department of Food Engineering and Process Management, Warsaw University of Life Science, 02-787 Warszawa, Poland; anna_kaminska1@sggw.edu.pl (A.K.-D.); michal.bialik@gmail.com (M.B.)

2 Department of Machinery Exploitation and Management of Production Processes, University of Life Sciences in Lublin, Głęboka 28, 20-950 Lublin, Poland; slawomir.kocira@up.lublin.pl

3 Department of Landscape Management, Faculty of Agriculture and Technology, University of South Bohemia in České Budějovice, Branišovská 1645/31a, 370 05 České Budějovice, Czech Republic

4 Department of Food Process Engineering, University of Life Sciences in Lublin, Doświadczalna 44, 20-280 Lublin, Poland; tomasz.oniszczuk@up.lublin.pl

5 Department of Physical Properties of Plant Materials, Institute of Agrophysics, Polish Academy of Sciences, 20-290 Lublin, Poland; m.stasiak@ipan.lublin.pl

* Correspondence: ewa_gondek@sggw.edu.pl

Abstract: Moldavian dragonhead leaves are a valuable waste product from the production of seed oil, from a nutritional point of view. It has been shown that dried leaves can be used for the production of functional foods. Despite the wide possibilities of using dried Moldavian dragonhead leaves, research on the selection of drying methods is scarce. This particular study investigated the traditional convection and the hybrid microwave–convection methods to dry Moldavian dragonhead leaves. We showed that the convection method supported by microwaves allowed us to obtain dried leaves in a much shorter time than traditional convection drying (the convection drying lasted, depending on the parameters, 184-75 min, while the convection–microwave drying lasted 33-7 min). Moreover, the microwave–convection-dried leaves contained more polyphenols (the average TPC in the convection droughts was 29.532, while in the microwave–convection droughts, it was 41.880 mg GAE/g d.m.) and chlorophylls (an average of 2.326 in the convection-dried and 2.675 mg/g d.m. in the microwave–convection-dried material) and were characterized by significantly higher antioxidant activity, compared to convection-dried ones.

Keywords: Moldavian dragonhead leaves; drying; microwave–convection; polyphenols; chlorophyll; anti-radical activity; ABTS; DPPH

1. Introduction

The Moldavian dragonhead (*Dracocephalum moldavica* L., *Lamiaceae*) is an annual herbaceous plant. It is mainly cultivated as an ornamental plant or melliferous plant.

Seeds are the most valuable part of the plant, from a nutritional point of view. The oil obtained from the seeds is rich in a large amount of omega-3 fatty acids, and the flowers are valuable in beekeeping. Moldavian dragonhead leaves are usually treated as a by-product, although many researchers already showed that the leaves contain ingredients that are valuable, from a pharmaceutical and nutritional point of view. The green parts of the Moldavian dragonhead contain mainly terpenoids (neral, geraniol, and geranyl acetate) and flavonoids (iridoids), as well as tannins [1–3]. The oil and extracts from the dried dragonhead leaves are used in the food industries [2,4,5]. Dried leaves are used as a tea in folk medicine for the treatment of coronary heart disorders and hypertension [6–8]. Recent research has shown that they can also be used as a valuable ingredient in functional foods. Wojtowicz et al. [9] showed that it is possible to produce corn crisps enriched with dried leaves. Dziki et al. [10] used dried Moldavian dragonhead leaves as an addition to bread. In both cases, the obtained product had favorable quality features and was highly accepted by consumers.

In order for the leaves of the Moldavian dragonhead to be used more widely for the production of functional food, a way to properly preserve them must be found.

Rudy et al. [2] examined the three most popular drying methods (convection, vacuum, and freeze drying) for drying Moldavian dragonhead leaves. The authors made an extensive assessment of the impact of the drying method on the biochemical properties of the dried leaves (Quantity and Composition of Essential Oils, Antiradical Activity, Total Flavonoids Content, Total Monomeric Anthocyanins, and Total Phenolic Compounds Content). They showed that in order to maintain the best quality, sublimation or vacuum methods should be used (at temperatures up to 40 °C), although these methods are time- and energy-consuming. The qualities of a dehydrated food product, like the flavors, content of heat- and oxygen-sensitive nutrients, product browning, rehydration capacity, appearance, and texture, depend on the duration and the temperature of the drying process [11–13]. Therefore, drying should take place under special conditions: the shortest possible duration and the lowest possible temperature. Freeze drying is carried out in a low-temperature chamber and without access to oxygen; it allows for the obtainment of a good-quality product, but it is a very energy-consuming and expensive process.

Recent research on the intensification of the drying process focuses on two directions: modern methods of pre-treatment and combining various drying techniques in one process [11,12].

The second-mentioned group includes microwave–convection drying, which is especially recommended for herbs [14]. The advantages of using microwaves to support drying include increasing the drying speed while limiting the heating of the material surface, reducing energy consumption, and improving the quality of the obtained dried products [14,15].

This is due to the mechanism of microwave heating, which is significantly different from convection heating. Microwaves penetrate the heated material and generate heat throughout its entire volume. The acceleration of heat and mass transfer during microwave drying results from the creation of internal pressure, which keeps the capillaries of the material open and allows water to move to the surface of the material [16,17].

Many researchers have used hybrid microwave–convection drying to dry leaves. Alibas [18] analyzed three methods of drying grape leaves: microwave, convection, and microwave–convection. She showed that the best method in terms of drying time, dried product color, and ascorbic acid content was the microwave–convection method. Akbudak and Akbudak [19] dried parsley leaves using convection, vacuum, and microwave methods. The criteria for the quality of the dried product were the color and general appearance of the dried product and the retention of ascorbic acid and chlorophylls. They indicated that the microwave method was optimal, and the microwave power in the range of 750–850 W allowed for the obtainment of the best-quality dried parsley leaves. Sarimsesli [20] dried coriander leaves using the microwave method and showed that the microwave drying process was faster compared to traditional techniques, and the dried material did not differ significantly in color from the fresh material. Arslan et al. [21] dried peppermint leaves by the convection, sun, and microwave methods and studied the kinetics of the process and the color, mineral content, and polyphenols in the dried leaves. They also indicated the microwave method as significantly shortening the drying time and providing the better retention of polyphenols. Yousif et al. [22] dried basil using traditional hot air and hybrid vacuum–microwave dryers; they studied the effect of the drying methods on the content of volatile flavor substances, the rate of rehydration, and the color and the structure of the basil leaves. The vacuum–microwave-dried basil contained more volatile substances than fresh basil. The air-dried basil leaves were darker and had a lower rehydration index than the microwave–vacuum-dried basil leaves. The same experimental design was used in the work by Yousif et al. [23], in which the research material was Mexican oregano leaves. In this case, the better retention of volatile components, color, and the better maintenance of structure were also observed in the case of using the hybrid method. Śledź et al. [14] studied the kinetics of the microwave–convection drying of basil (*Ocimum basilicum*), lovage (*Levisticum officinale*), mint (*Mentha* sp.), oregano (*Origanum vulgare*), parsley (*Petroselinum crispum*), and rocket (*Eruca vesicaria*) using different mathematical models. They also analyzed changes in the content of polyphenols, color, and sorption properties of the dried materials. They stated that the highest retention of phenolic compounds, the best color stability, and the highest water vapor sorption capacity were characterized by lovage and parsley.

In recent years, many studies have been conducted using the hybrid microwave–convection drying of food raw materials other than herbs. Krzykowski et al. [24] examined the suitability of different drying methods (freeze drying, convection, vacuum, and hybrid microwave–convection) for drying red peppers. These authors showed that freeze drying and vacuum drying cause the smallest losses of bioactive substances, while convection–microwave drying is the fastest among the tested methods of removing water from peppers. Similar conclusions were obtained by Garcia-Moreira et al. [25], who compared the open-air, solar, microwave, freeze drying, and shade drying of red peppers. They found that the use of microwave assistance significantly shortens the drying process and the combination of air drying at a temperature of 60 °C and microwaving at a power of 160 W results in the best retention of the components responsible for the antioxidant activity of peppers.

Luka et al. [26] compared the convection and the microwave drying of white cabbage (*Brassica oleracae* var *capitata* L.). Kaur et al. [27] investigated different drying and processing methods to obtain good-quality powder from banana peels. It was shown that microwave–convection drying, preceded by ultrasonic treatment, is optimal.

Inla et al. [28] compared the properties of hybrid convection–microwave-dried lingzhi mushroom (*Ganoderma lucidum*). Behera and Balasubramanian [29] compared the convection and the microwave–convection methods for drying microalgae, and showed that, in addition to shortening the drying time, the use of microwaves causes a higher lipid yield and the better preservation of bioactive components.

All the mentioned researchers who used the hybrid microwave–convection drying method emphasize the beneficial combination of convection drying with microwave heating. These processes complement each other in terms of providing heat to the dried material: hot air heats the material's surfaces, while microwave power heats the material from the inside. This combination causes faster water evaporation, which takes place simultaneously by means of liquid water diffusion, water vapor diffusion, and the capillary and hydrodynamic water flow in the material [18]. Reducing the drying time, in turn, shortens the time of the exposure of the material to high temperatures and atmospheric oxygen, which can contribute to the better retention of thermolabile components and the retention of the material's color and structure [14,17,19–29]. The aim of this study was to investigate whether hybrid microwave–convection drying is a suitable method for drying Moldavian dragonhead leaves and how it affects the bioactive components that determine the health-promoting effects of this herb.

2. Materials and Methods

2.1. Material

Moldavian dragonhead (*Dracocephalum moldavica* L.), with blue flowers, was cultivated on a 0.5-hectare plot in Czesławice (51°18′ N; 22°15′ E). The experiment was conducted on medium soil with a sandy silt texture and a pH of 7.4–7.5. Phosphorus, potassium, and magnesium levels were high. The cultivation followed integrated production standards. Before sowing, mineral fertilization was applied at 30 kg N/ha, 60 kg P/ha, and 90 kg K/ha. After emergence, at the BBCH 12 stage, nitrogen was applied at 20 kg N/ha. *Dracocephalum moldavica* seeds were sown on 2 May 2019, with a mechanical point seeder at a spacing of 45 cm × 15 cm and a depth of 1–1.5 cm. Meteorological conditions during the growing season were favorable, with average monthly temperatures from May to July ranging from 12.6 to 19.4 °C and rainfall from 74 to 173 mm. The plants for further research were harvested at the BBCH 61 stage.

2.2. Drying

Convection drying (CD) was carried out in a prototype laboratory dryer at the Department of Food Engineering and Production Organization of the Institute of Food Sciences of the Warsaw University of Life Sciences (Figure 1). The leaves (approximately 20 g) were placed on a sieve with a load of 0.7 kg/m^2, and each time, they occupied the same area. The air flow was 0.5 m/s and was parallel to the samples. The ambient air humidity was approximately 55%. The drying was carried out until the water content was below 0.1 kg/kg d.m. The drying was carried out at the following temperatures: 40–50, 60, 70 °C. The kinetics of the drying process were examined by interval measurement and the recording of the mass of the tested samples in the computer memory. Each drying was performed in 2 repetitions.

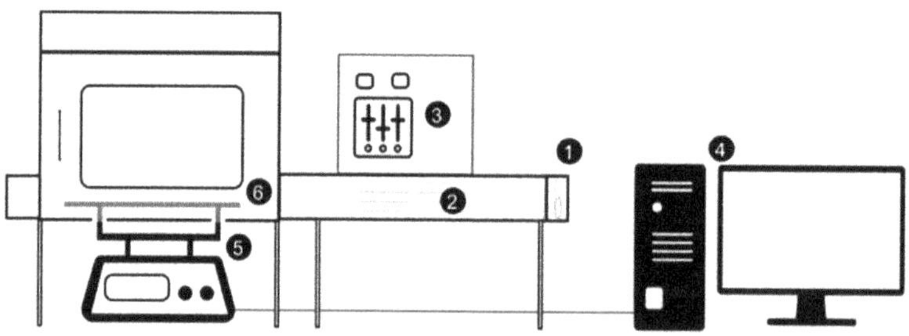

Figure 1. Scheme of a convection dryer: 1—fans; 2—heating element; 3—control section; 4—recording computer; 5—balance; 6—drying chamber.

Microwave–convection drying (M-C) was carried out in a prototype laboratory dryer at the Department of Food Engineering and Production Organization, the Institute of Food Sciences, the Warsaw University of Life Sciences (Figure 2). The leaves were dried on a rotating sieve with a load of 0.7 kg/m^2, and the material occupied the same surface each time. The air flow was 0.5 m/s and was perpendicular to the samples; the microwave power (with a typical frequency of 2450 MHz) was 100, 200, and 300 W, respectively; and the temperature of the drying air was at the level of 30 and 40 °C. The drying was carried out until the water content was below 0.1 kg/kg d.m. The dryer allowed for interval measurement and the recording of the mass of the tested samples in the computer memory. The drying was performed in two repetitions.

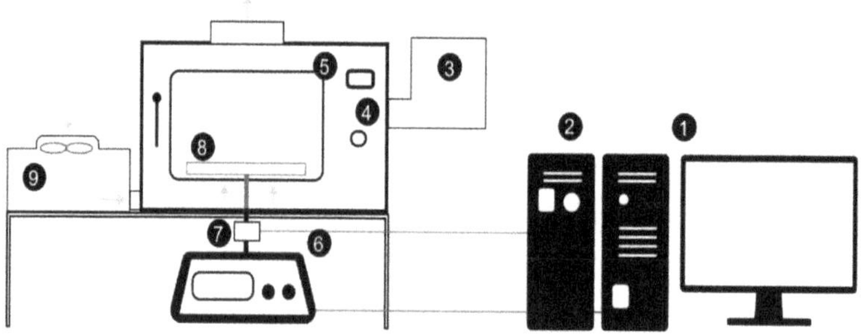

Figure 2. Scheme of a microwave–convection dryer: 1. recording computer; 2. control computer; 3. magnetron; 4. air flow control; 5. drying chamber; 6. scale; 7. the rotating mechanism of the drying sieve; 8. perforated sieve; 9. Fan.

The drying curves were plotted as a function of dimensionless moisture ratio (MR) during drying. The moisture ratio was calculated as follows:

$$MR = (u_\tau - u_e)/(u_0 - u_e) \quad (1)$$

where u_e is the equilibrium moisture content (g $H_2O \cdot g^{-1}$ dm), u_0 is the initial moisture content (g $H_2O \cdot g^{-1}$ dm), and u_τ is the moisture content at each moment of the process (g $H_2O \cdot g^{-1}$ dm).

Table Curve 2D v5.01 software was used to compute the effective water diffusion coefficient. The simplified equation of Fick's second law of diffusion (Table 1) for an infinite plate was used [30].

$$MR = (8/\pi^2) \cdot \exp((-\pi^2 D_{eff} \tau)/(4L^2)) \quad (2)$$

where τ—time (s); D_{eff}—effective water diffusion coefficient ($m^2 \cdot s^{-1}$); and L—half of the material thickness (m).

Table 1. Effective moisture diffusion coefficient and time of drying.

Drying Method	$D_{eff}(m^2 \cdot s^{-1})$	Drying Time, min	Water Activity
CD 40C	$7.103 \cdot \times 10^{-12}$	184	0.962 ± 0.023 a
CD 50C	$9.522 \times \cdot 10^{-12}$	134	0.183 ± 0.013 c
CD 60C	$1.503 \times \cdot 10^{-11}$	105	0.170 ± 0.021 c
CD 70C	$9.103 \times \cdot 10^{-11}$	75	0.103 ± 0.020 d
MW-CD 100W 30C	$9.783 \cdot \times 10^{-11}$	33	0.081 ± 0.003 d
MW-CD 100W 40C	$1.019 \times \cdot 10^{-10}$	18	0.214 ± 0.011 b
MW-CD 200W 30C	$3.477 \cdot \times 10^{-10}$	12	0.218 ± 0.014 b
MW-CD 200W 40C	$8.571 \times \cdot 10^{-10}$	11	0.201 ± 0.009 b
MW-CD 300W 30C	$1.004 \cdot \times 10^{-9}$	9	0.165 ± 0.017 c
MW-CD 300W 40C	$1.103 \times \cdot 10^{-9}$	7	0.199 ± 0.011 c

Mean values followed by the same small letter (in columns) do not differ significantly at α = 0.05.

2.3. Water Activity

Water activity (aw) was measured using a HygroLab DT2 ROTRONIC (Rotronic AG, Bassersdorf, Switzerland) apparatus at a temperature of 23 ± 1 °C. The measurement was carried out in 5 repetitions.

2.4. Total Phenolic Content

The extraction of phenolics from a dried, powdered Moldavian dragonhead leaf 80% ethanol solution was carried out in two independent extracts, prepared from each variant of the experiment, as already described by Bialik et al. [31]. The total phenolic content was based on a Folin–Ciocalteau methodology, proposed by Singleton and Rossi [32] with some modifications [33]. This procedure was conducted as follows: First, 30 mL of distilled water, 1.5 mL of extract, and 3 mL of Folin–Ciocalteau reagents were mixed. After 3 min of incubation, 2 mL sodium carbonate was added, and the solution was stirred. Afterwards, the tubes with the prepared samples were filled with distilled water up to 50 mL, mixed, and kept in darkness for 1 h. The absorbance was checked against a blank sample (prepared in an analogous way, but without any extract, just distillated water) at 750 nm (Helios v. 7.03 spectrophotometer, Thermo Electron Corporation, Waltham, MA, USA). The examined total phenolic content was expressed as the mg of gallic acid equivalent per gram of dry mass [mg GAE/g d.m.] [31].

2.5. Chlorophyll a and b Content

The content of chlorophyll a and b was determined based on a method that was shown by śledz et al. [34]. Approximately 0.1 g of dried Moldavian dragonhead leaves was weighed with an accuracy of ±0.0001 g into a 50 mL centrifuge tube. To facilitate the formation of a sediment during centrifugation, 0.1 g of $MgCO_3$ was added. The contents of the test tubes prepared in this way were poured into 20 mL of 80% acetone. Then the samples were homogenized in an IKA rotor-stator homogenizer, Yellowline D125 Basic, for 1 min at 13,500 rpm. The homogenizing tip was washed with 80% acetone to recover the particles that remained there. The prepared test tubes were mixed in a Vortex shaker at 2400 rpm for 2 min. The pellet was centrifuged for 5 min in a Sigma Laboratory Centrifuge (VWR, Leuven, Belgium). The supernatant from the sediment was poured into previously prepared volumetric flasks 25 mL and topped up with 80% acetone. The extraction process was performed in 4 repetitions for each material. The absorbance was measured in a Heλios γ Thermo Spectronic spectrophotometer (Waltham, MA, USA) against a blank sample consisting of acetone at a concentration of 80% at wavelengths of 470, 646 and 663 nm [31].

2.6. ABTS Analysis

The experiments were performed according to [31], with small modifications. The antioxidant properties were measured according to the method of Re et al. [35], using the ABTS assay method. The ABTS reagent was composed by mixing 3.84 mM of ABTS with 0.002 mM of potassium persulfate in water and, afterwards, incubating in the dark for >16 h. The concentrated ABTS reagent was then diluted with phosphate-buffered saline to adjust a final absorbance at the level of 0.72 (±0.2) at 734 nm. Ultimately, 3 mL of ABTS solution was added to four Moldavian dragonhead leaf extracts, ranging from 0.02 to 0.04 mL (as filled up to 2 mL of ethanol). Each sample absorbance was measured at a wavelength of 734 nm in two runs after 10 min in a spectrophotometer, Helios v. 7.03 (Waltham, MA, USA), against a blank sample—ethanol solution. The results were expressed as EC50 ABTS (the concentration of the dry matter of the extracts needed to scavenge 50% of the initial ABTS radicals). The measurement was performed in two replicates for each individual extract.

2.7. DPPH Analysis

The radical scavenging activity was measured by assessing the degree to which DPPH radicals were neutralized by the antioxidants of the Moldavian dragonhead leaf extracts, as specified by Brand-Williams et al. [36]. (Extracts ranging from 0.05 to 0.35 mL (6 samples) were filled up to 2 mL using an 80% aqueous solution of ethanol, mixed with 2 mL of DPPH 0.22 mM solution. Afterward, the absorbance was measured after 30 min of incubation in darkness at a wavelength of 517 nm (Helios v. 7.03 spectrophotometer—Waltham, MA, USA) against the blank—an 80% (v/v) ethanol solution. The results were expressed as the concentration of the dry matter of the extracts needed to scavenge 50% of the initial DPPH radicals (EC50 DPPH). The measurements were repeated in two replicates for each extract [31].

2.8. Statistical Analysis

The statistical analysis was performed using Statistica 12 (StatSoft) and Excel 2013 (Microsoft) software. The analysis of variance (ANOVA) with a post-hoc Tukey's test at significance level of $\alpha = 0.05$ was used.

3. Results and Discussion

3.1. Drying

Traditional convection drying and hybrid microwave–convection drying were analyzed. Figure 3 shows the changes in the reduced water content (MR) as a function of the duration of the convection drying and the microwave–convection drying of the Moldavian dragonhead leaves. In the case of convection drying, a reduction in the drying time was observed with increasing temperature. The convection drying time ranged from 184 min at 40 °C to 75 min at 70 °C and was similar to that obtained in the work of Rudy et al. [2]. The microwave support resulted in a radical shortening of the process. The microwave drying time ranged from 33 min (MW-CD 100W 30C) to 7 min (MW-CD 300W) (Table 1).

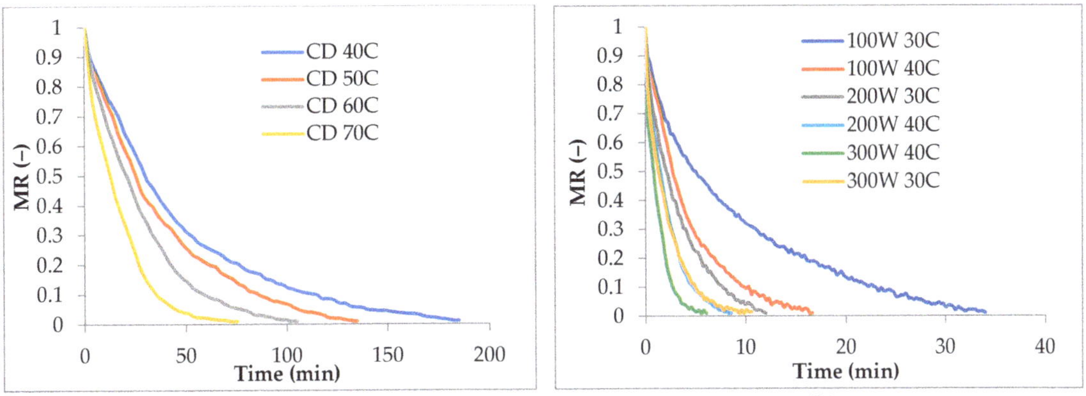

(**a**) (**b**)

Figure 3. Drying curves: (**a**) convection drying; (**b**) microwave–convection drying.

The combined microwave–convection drying time depended on both the air temperature and the microwave power. Similar observations were made by many authors using microwave-assisted drying [17–20,33,34,37]. During microwave drying, the drying speed decreases as the amount of water dipoles decreases. In Figure 3 this phenomenon was visible in the drying curves that correspond to higher microwave power (200 and 300 W). Similar results were obtained in the case of drying many leaf materials [22,29,30]. In some works, the opposite effect was observed, consisting of a sudden acceleration of the drying rate at MR = 0.4. This is explained by the formation of stresses causing tissue disruption and the leakage of cellular contents [29,38]. The intensity of the drying process can also be analyzed based on the values of the effective water diffusion coefficient. Microwave radiation increases the intensity of the process, which is reflected in the increasing D_{eff} coefficient. The water diffusion coefficients obtained in this study are presented in Table 1. In the case of convection drying, the D_{eff} was in the range of $7.103 \cdot 10^{-12}$–$9.103 \cdot 10^{-11}$, and in the case of microwave–convection drying, the D_{eff} was in the range of $9.783 \cdot 10^{-11}$–$1.103 \cdot 10^{-9}$ $m^2 \cdot s^{-1}$. The increases in both the temperature and the microwave power resulted in an increase in the D_{eff} values (Table 1). The obtained dried samples slightly differed in water activity, corresponding to the same water content (MR).

Similarly, low D_{eff} values, obtained during the microwave–convection drying of leaves, were observed by other researchers [25,38,39].

3.2. Chlorophylls and Polyphenols Content

The chlorophyll content of the leaves of Moldavian dragonhead varies depending on the region in which it naturally grows [40]. It was shown that the amounts of chlorophyll a, b, and the total chlorophyll are affected by the factors of genotype and fertilization regime [41]. The chlorophyll a and b content in the fresh leaves was 5.362 ± 0.473 mg/g d.m. The application of convection and microwave–convection drying caused a significant decrease in the content of both chlorophyll (a and b) pigments (Table 2).

After the process, the content of chlorophyll a and b varied from 2.199 ± 0.277 to 3.124 ± 0.224 mg/g d.m. The statistical analysis showed that the convection method caused higher chlorophyll losses compared to the convection–microwave method. Only the sample dried at the highest microwave power (MW-CD 300W) and at the temperature of 40 °C formed a homogeneous group with the samples dried by the convection method. The content of chlorophylls in the samples dried by microwave–convection was influenced by both the temperature and the microwave power. The highest content of chlorophyll a and b was noted in the sample dried by microwave–convection at 30 °C and a microwave power of 100 W, the lowest was noted in the material dried by convection at 70 °C and by microwave–convection at 40 °C and a microwave power of 200 W.

Table 2. Polyphenol and chlorophyll content in fresh and dried Moldavian dragonhead leaves.

Drying Method	TPC (mg GAE/g d.m.)	Chlorophyll a and b Content (mg/g d.m.)
Fresh	89.087 ± 8.515 a	5.362 ± 0.473 a
CD 40C	32.262 ± 0.094 d	2.499 ± 0.277 de
CD 50C	33.676 ± 1.229 d	2.311 ± 0.085 e
CD 60C	28.701 ± 2.191 de	2.362 ± 0.347 de
CD 70C	23.489 ± 1.683	2.132 ± 0.607 e
MW-CD 100W 30C	46.298 ± 0.808 b	3.124 ± 0.224 b
MW-CD 100W 40C	41.904 ± 2.857 c	2.970 ± 0.138 c
MW-CD 200W 30C	44.246 ± 0.744 bc	2.940 ± 0.137 c
MW-CD 200W 40C	42.652 ± 2.566 bc	2.619 ± 0.254 cd
MW-CD 300W 30C	39.173 ± 2.035 cd	2.479 ± 0.083 d
MW-CD 300W 40C	41.427 ± 0.091 c	2.365 ± 0.031 e

Mean values followed by the same small letter (in columns) do not differ significantly at $\alpha = 0.05$.

Śledź and Witrowa-Rajchert [42] analyzed the chlorophyll content in microwave–convection-dried herbs (basil, lovage, mint, oregano, parsley, and rocket leaves). These authors also found chlorophyll loss during the drying process; they showed that chlorophyll retention during drying is largely dependent on the type of plant: the highest was in the case of basil and lovage (over 90%), and the lowest was observed in rocket (32–25%). Di Cesare et al. [43,44] compared the content of bioactive compounds in basil dried using different methods. They also found that chlorophyll retention in microwave–convection-dried samples is higher than in convection-dried ones. Similar relationships were observed by other authors [19,20,43–46].

The content of polyphenols in the green parts of plants is very diverse and largely dependent on the species. The green parts of Moldavian dragonhead are reported to contain numerous polyphenols, especially hydroxycinnamic acids (rosmarinic and caffeic acids) and flavonoids such as apigenin and luteolin and their glycosides, quercetin, diosmetin, kaempferol, acacetin, agastachioside, and salvigenin [2,7]. Investigations have revealed significant therapeutic effects of D. moldavica extracts: cardioprotective, sedative, antiplatelet, and antiaging [2,7,47].

Phenolic compounds have structures that are very easily oxidized during the processing of plant food. Oxidation can lead to the reduction of radical scavenging capacities. The factors that accompany the drying process, such as the effect of high temperatures and oxygen in the atmosphere for a long time, are the most important factors that affect the total polyphenol content. It has been shown that among the technological factors that accompany food processing, the combination of temperature and oxygen has a particularly adverse effect on phenolic compounds [45]. Sweeney and Seetharaman [45] presented an overview of drying technologies and their impact on the polyphenol content of vegetables and fruits.

They indicated microwave drying as one of the drying methods that leads to reducing polyphenol losses.

Table 2 presents the content of polyphenols in the fresh leaves and in the dried material. The content of polyphenols in the fresh leaves was similar to that obtained by other authors [2]. The drying process, in the case of both analyzed methods, led to a decrease in the content of polyphenols. The highest loss was observed in the case of convection drying at the temperature of 70 °C (a decrease of 75%). In convection drying, materials were used at the temperature of 40–60 °C; the losses of polyphenols were also significant, higher than in the work of Rudy et al. [2]. High losses of polyphenols at low drying temperatures result from the extension of the drying time, with a decrease in the temperature of the drying agent. This causes a longer duration of the exposure of the material to atmospheric oxygen. Rudy et al. [2] studied the effect of the drying method and temperature on the retention of polyphenols and indicated the convection method as the one in which the losses are the highest. They found that the vacuum and freeze drying methods are better than convection in terms of preserving the biological activity of the dried material.

The higher retention of polyphenols in microwave–convection-dried leaves has been noted by many authors. This is most often explained by a shorter time of exposure of the material to oxygen and higher temperatures. However, it is also possible that microwaves can cause the destruction of plant cells, which increases the extractability of polyphenols.

Śledź et al. [14] noticed that the degradation of polyphenols in microwave–convection-dried herbs was different in different herbs. Drying caused a significant degradation of polyphenols in mint, oregano, and basil (56, 40, and 34%, respectively, compared with fresh leaves). In turn, lovage, parsley, and rocket leaves were not influenced significantly by the total phenolic content.

Among the analyzed variants of microwave–convection drying, the best retention of polyphenols was found in dried leaves in which the microwave power was 100 and 200 W (Table 2). Increasing the microwave power to 300 W caused a decrease in the content of polyphenols in dried leaves of Moldavian dragonhead, which could be the result of the material overheating.

3.3. Antioxidant Activity

Free radicals are groups of atoms or molecules that have one or more unpaired electron, which makes them chemically unstable and highly reactive. The presence of free radicals in the body is associated with the development of many diseases and also accelerates the aging of tissues. An important role in preventing the effects of free radical activity is played by antioxidants, which are supplied to the human body in food. The most important source of antioxidants in the diet are fruits, vegetables, and herbs. However, the preservation of these products is associated with significant losses of valuable antioxidants. Drying is the most popular method of preserving herbaceous plants; therefore, the search for the method and the parameters of drying leading to the preservation of antioxidant capacity is very important [46,48,49].

Antioxidant activity is caused by numerous agents, such as vitamins (A, C, and E), minerals (zinc, selenium, copper, manganese, and iron), carotenoids (α- and β-carotene, zeaxanthin, lutein, lycopene, and β-cryptoxanthin), and polyphenols (flavonoids, phenolic acids, lignans, and stilbenes). Most of them are sensitive to high temperatures. There are many methods for testing antiradical activity; in this work antioxidant activity was analyzed based on the scavenging of two synthetic radicals, DPPH* and $ABTS^{\circ+}$. The results were expressed as EC_{50}, i.e., the concentration of the leaf extract (fresh and dried) required to inhibit half of the initial amount of the radical. This means that lower EC_{50} values indicate higher antiradical activity [17,41,46,48].

The drying of Moldavian dragonhead leaves, regardless of the parameters used, caused a decrease in the radical scavenging of the dried material, compared to the raw material (Figure 4).

The EC_{50} $ABTS^{\circ+}$ of the extracts from the fresh leaves was 11.20 mg chlorogenic acid/g d.m., while in the convection-dried leaves it was 19.61–24.5 mg chlorogenic acid/g d.m. and in the microwave-dried leaves it was 16.5–19.9 mg chlorogenic acid/g d.m. The EC_{50} DPPH* in the fresh leaves was 14.96 mg TE/g d.m., while in the convection-dried leaves it was 23.23–27.41 and in the microwave–convection-dried leaves it was 18.12–25.14 mg TE/g d.m. A decrease in radical scavenging activity during the drying of herbaceous plants was found in many works [2,14,46–48].

Similar results were obtained by Rudy et al. [2]. The researchers analyzed the antioxidant activity of fresh and convection-, vacuum-, and freeze-dried leaves of Moldavian dragonhead. The highest radical scavenging activity was characterized by freeze drying, the lowest by convection drying, while the increase in the temperature of the drying agent decreased the antioxidant activity measured by $ABTS^{\circ+}$ [2].

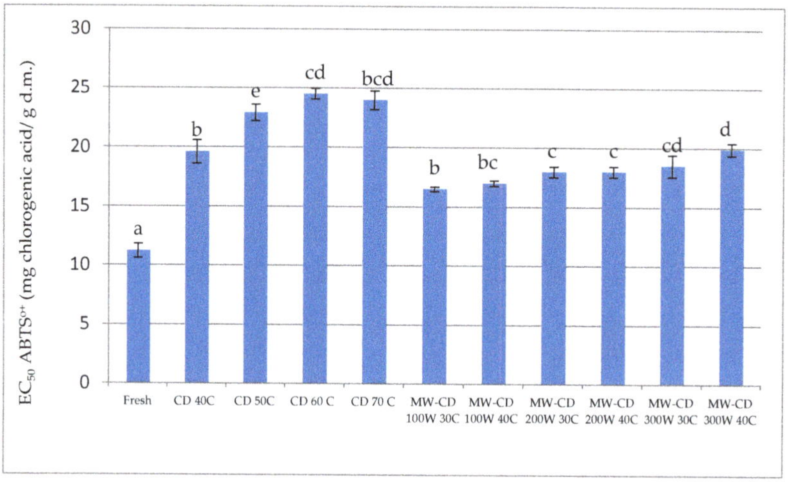

Figure 4. ABTS free radical scavenging ability in fresh and dried Moldavian dragonhead leaves. Mean values followed by the same small letter do not differ significantly at $\alpha = 0.05$.

In this study, a decrease in antioxidant activity was also observed with an increase in the temperature of the drying agent. The leaves dried at 40 °C were characterized by an antiradical activity of 19.6 mg chlorogenic acid/g d.m. EC_{50} $ABTS^{\circ+}$ and 14.96 mg TE/g d.m EC_{50} DPPH*, while those dried at 70 °C were characterized by an antiradical activity of 23.99 mg chlorogenic acid/g d.m. and 27.41 mg TE/g d.m (Figure 5). The correlation coefficient between TPC and $DPPH^*$ was −0.89, and between TPC and $ABTS^{\circ+}$ it was −0.88. This means that the antioxidant activity of Moldavian dragonhead leaves is created mainly by phenolic compounds.

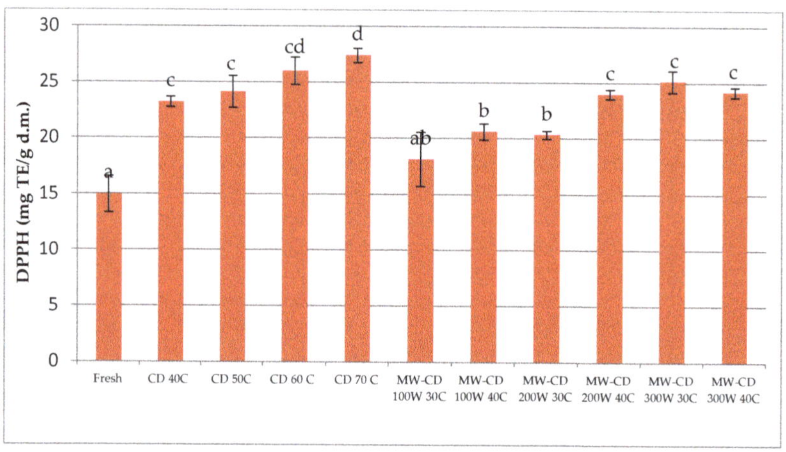

Figure 5. DPPH free radical scavenging ability of fresh and dried Moldavian dragonhead leaves. Mean values followed by the same small letter do not differ significantly at $\alpha = 0.05$.

Despite the fact that the dried leaves of Moldavian Dragonhead are a well-known and valued element of folk medicine, the scientific literature available so far lacks extensive information on the selection of appropriate drying methods. The studies published so far by Rudy et al. [2] concerned several selected conventional drying methods. The demonstration that the hybrid microwave–convection drying method allows for a significant reduction in drying time compared to conventional methods while maintaining the good retention of thermolabile components is an important conclusion resulting from the presented studies. Further work will be aimed at the optimization of the microwave–convection drying parameters and the assessment of the energy aspects of the process. The results obtained on a laboratory scale will also require experiments that will allow the developed methodology to be transferred to an industrial scale, which may cause some difficulties.

4. Conclusions

The use of a hybrid microwave–convection method for drying Moldavian dragonhead leaves allows for a significant reduction in the drying time compared to traditional drying using hot air. The microwave–convection drying time depends on both the time and the air temperature.

The content of biologically active compounds, namely, polyphenols and chlorophylls a and b, was significantly dependent on the drying temperature and was, regardless of the parameters used, higher in the microwave–convection-dried materials. This was probably due to the shorter duration of the effect of high temperatures and atmospheric oxygen on the dried material during microwave-assisted drying.

Drying, in all the tested methods and parameters, resulted in a decrease in the ability to capture free radicals compared to fresh leaves. However, the microwave–convection-dried materials were characterized by higher antiradical activity than convection-dried materials. A high correlation was found between antiradical activity and TPC, which allows us to assume that polyphenols are mainly responsible for the high antioxidant activity of Moldavian dragonhead leaves.

Author Contributions: Conceptualization, E.G., A.K.-D. and S.K.; methodology, M.B., E.G. and A.K.-D.; validation, T.O. and M.S.; formal analysis, M.S., T.O. and A.K.-D.; investigation, E.G., M.B., A.K.-D., T.O., M.S. and S.K.; writing—original draft preparation, E.G. and A.K.-D.; writing—review and editing, E.G., A.K.-D., M.B., M.S., T.O. and S.K.; supervision, E.G., A.K.-D. and M.S. All authors have read and agreed to the published version of the manuscript.

Funding: This research received no external funding.

Institutional Review Board Statement: Not applicable.

Informed Consent Statement: Not applicable.

Data Availability Statement: The data generated or analyzed during this study are available from the corresponding author on reasonable request.

Acknowledgments: Special thanks to Magdalena Barwińska for the help with the experiment's implementation.

Conflicts of Interest: The authors declare no conflicts of interest.

Nomenclature

a, b, c, d, e…	homogenous group
CD	convection drying
MW—CD	microwave–convection drying
D_{eff}	effective water diffusion coefficient ($m^2 \cdot s^{-1}$)
L	half of the material thickness (m)
MR	moisture ratio
τ	time (min)
u_e	the equilibrium moisture content (g $H_2O \cdot g^{-1}$ dm)
u_0	the initial moisture content (g $H_2O \cdot g^{-1}$ dm)
u_τ	the moisture content at each moment of the process (g $H_2O \cdot g^{-1}$ dm)

References

1. Sultan, A.; Bahang, A.; Aisa, H.A.; Eshbakova, K.A. Flavonoids from *Dracocephalum moldavica*. *Chem. Nat. Compd.* **2008**, *44*, 336–337. [CrossRef]
2. Rudy, S.; Dziki, D.; Biernacka, B.; Krzykowski, A.; Rudy, M.; Gawlik-Dziki, U.; Kachel, M. Drying Characteristics of *Dracocephalum moldavica* Leaves: Drying Kinetics and Physicochemical Properties. *Processes* **2020**, *8*, 509. [CrossRef]
3. Janmohammadi, M.; Youefzadeh, S.; Sabaghnia, N. Chemical comparison of essential oils in dragonhead (*Dracocephalum moldavica* L.) samples grown in different areas. *J. Essent. Oil-Bear Plant* **2018**, *21*, 950–962. [CrossRef]
4. Dmitruk, M.; Weryszko-Chmielewska, E. Morphological differentiationand distribution of non-glandular and glandular trichomes on *Dracocephalum moldavicum* L. shoots. *Acta Agrobot.* **2010**, *63*, 97–110. [CrossRef]
5. Yousefzadeh, S.; Modarres-Sanavy, S.A.M.; Sefidkon, F.; Asgarzadeh, A.; Ghalavand, A.; Sadat-Asilan, K. Effects of Azocompostandurea on the herbage yield and contents and compositions of essentialoils from two genotypes of dragonhead (*Dracocephalum moldavica* L.) in two regions of Iran. *Food Chem.* **2013**, *138*, 1407–1413. [CrossRef]
6. Maimaitiyiming, D.; Hu, G.; Aikemu, A.; Hui, S.W.; Zhang, X. The treatment of Uygur medicine *Dracocephalum moldavica* L. on chronic mountains sickness rat model. *Pharmacogn. Mag.* **2014**, *10*, 477–482. [CrossRef]
7. Yang, L.N.; Xing, J.G.; He, C.H.; Wu, T. The phenolic compounds from *Dracocephalum moldavica* L. *Biochem. Syst. Ecol.* **2014**, *54*, 19–22. [CrossRef]
8. Omidbaigi, R.; Yavari, S.; Hassani, M.E.; Yavari, S. Induction of autotetraploidy in dragonhead (*Dracocephalum moldavica*) by colchicine treatment. *J. Fruit. Ornam. Plant Res.* **2010**, *18*, 23–35.
9. Wójtowicz, A.; Oniszczuk, A.; Oniszczuk, T.; Kocira, S.; Wojtunik, K.; Mitrus, M.; Kocira, A.; Widelski, J.; Skalicka-Woźniak, K. Application of Moldavian dragonhead (*Dracocephalum moldavica* L.) leaves addition as a functional component of nutritionally valuable corn snacks. *J. Food Sci. Tech. Mys.* **2017**, *54*, 3218–3229. [CrossRef]
10. Dziki, D.; Cacak-Pietrzak, G.; Gawlik-Dziki, U.; Sułek, A.; Kocira, S.; Biernacka, B. Effect of moldavian dragonhead (*Dracocephalum moldavica* L.) leaves on the baking properties of wheat flour and quality of bread. *CyTA-J. Food* **2019**, *17*, 536–543. [CrossRef]
11. Bialik, M.; Wiktor, A.; Gondek, E. Advanced fruit drying technologies. *Adv. Agric. Sci. Probl.* **2019**, *597*, 15–22. [CrossRef]
12. Witrowa-Rajchert, D.; Wiktor, A.; Sledz, M.; Nowacka, M. Selected emerging technologies to enhance the drying process. A review. *Dry. Technol.* **2014**, *32*, 1386–1396. [CrossRef]
13. Zhang, M.; Chen, H.; Mujumdar, A.S.; Tang, J.; Miao, S.; Wang, Y. Recent developments in high-quality drying of vegetables, fruits, and aquatic products. *Crit. Rev. Food Sci. Nutr.* **2017**, *57*, 1239–1255. [CrossRef]
14. Sledz, M.; Nowacka, M.; Wiktor, A.; Witrowa-Rajchert, D. Selected chemical and physicochemical properties of microwave-convective dried herbs. *Food Bioprod. Process.* **2014**, *91*, 421–428. [CrossRef]
15. Nowacka, M.; Sledz, A.; Wiktor, D.; Witrowa-Rajchert, D. Physical and chemical properties od microwave dried food products. *Żywność Nauka Technologia Jakość* **2012**, *85*, 5–20. [CrossRef]
16. Moses, A.; Norton, T.; Alagusundaram, K.; Tiwari, B.K. Novel Drying Techniques for the Food Industry. *Food Eng. Rev.* **2014**, *6*, 43–55. [CrossRef]
17. Brewer, M. Microwave processing, nutritional and sensory quality. In *The Microwave Processing of Foods*; Schubert, H., Regier, M., Eds.; Woodhead Publishing Ltd.: Cambridge, UK, 2005; Available online: https://shop.elsevier.com/books/the-microwave-processing-of-foods/regier/978-0-08-100528-6 (accessed on 24 October 2024).
18. Alibas, I. Microwave, air and combined microwave-air drying of grape leaves (*Vitis cinifera* L.) and the determination of some quality parameters. *Int. J. Food Eng.* **2014**, *10*, 69–88. [CrossRef]
19. Alibas, I.; Köksal, N. Convective, vacuum and microwave drying kinetics of mallow leaves and comparison of color and ascorbic acid values of three drying methods. *Food Sci. Technol. Camp.* **2014**, *34*, 358–364. [CrossRef]
20. Sarimeseli, A. Microwave drying characteristics of coriander (*Coriandrum sativum* L.) leaves. *Energy Convers. Manag.* **2011**, *52*, 1449–1453. [CrossRef]
21. Arslan, N.D.; Özcan, M.M.; Mengeş, H.O. Evaluation of drying methods with respect to drying parameters, some nutritional and colour characteristics of peppermint (*Mentha × piperita* L.). *Energy Convers. Manag.* **2010**, *51*, 2769–2775. [CrossRef]
22. Yousif, A.N.; Scaman, C.H.; Durance, T.D.; Girard, B. Flavor volatiles and physical properties of vacuum-microwave-and air-dried sweet basil (*Ocimum basilicum* L.). *J. Agric. Food Chem.* **1999**, *47*, 4777–4781. [CrossRef] [PubMed]
23. Yousif, A.N.; Durance, T.D.; Scaman, C.H.; Girard, B. Handspace volatiles and physical characteristics of vacuum-microwave, air and freeze-dried oregano (*Lippia berlandieri* Schauer). *J. Food Sci.* **2000**, *65*, 926–930. [CrossRef]
24. Krzykowski, A.; Rudy, S.; Polak, R.; Biernacka, B.; Krajewska, A.; Janiszewska-Turak, E.; Kowalska, I.; Żuchowski, J.; Skalski, B.; Dziki, D. Drying of Red Chili Pepper (*Capsicum annuum* L.): Process Kinetics, Color Changes, Carotenoid Content and Phenolic Profile. *Molecules* **2024**, *29*, 5164. [CrossRef] [PubMed]
25. García-Moreira, D.P.; Pacheco, N.; Hernández-Guzmán, H.; Bahammou, Y.; Tagnamas, Z.; Moreno, I.; López-Vidaña, E.C. Evaluation of Various Drying Methods for Mexican *Yahualica chili*: Drying Characteristics and Quality Assessment. *Processes* **2024**, *12*, 1969. [CrossRef]
26. Luka, B.S.; Vihikwagh, Q.M.; Ngabea, S.A.; Mactony, M.J.; Zakka, R.; Yuguda, T.K.; Adnouni, M. Convective and microwave drying kinetics of white cabbage (*Brassica oleracae* var capitata L.): Mathematical modelling, thermodynamic properties, energy consumption and reconstitution kinetics. *J. Agric. Food Res.* **2023**, *12*, 100605. [CrossRef]

27. Kaur, M.; Modi, V.K.; Sharma, H.K. Effect of carbonation and ultrasonication assisted hybrid drying techniques on physical properties, sorption isotherms and glass transition temperature of banana (Musa) peel powder. *Powder Technol.* **2022**, *396*, 519–534. [CrossRef]
28. Inla, K.; Bunchan, S.; Krittacom, B.; Luampon, R. Drying behavior, color change and rehydration of lingzhi mushroom (*Ganoderma lucidum*) under convection-assisted microwave drying. *Case Stud. Therm. Eng.* **2023**, *49*, 103348. [CrossRef]
29. Behera, B.; Paramasivan Balasubramanian, P. Experimental and modelling studies of convective and microwave drying kinetics for microalgae. *Bioresour. Technol.* **2021**, *340*, 125721. [CrossRef]
30. Crank, J. *The Mathematics of Diffusion*, 2nd ed.; Oxford Calendon Press: Oxford, UK, 1975. Available online: https://www-eng.lbl.gov/~shuman/NEXT/MATERIALS&COMPONENTS/Xe_damage/Crank-The-Mathematics-of-Diffusion.pdf (accessed on 24 October 2024).
31. Bialik, M.; Wiktor, A.; Rybak, K.; Witrowa-Rajchert, D.; Latocha, P.; Gondek, E. The impact of vacuum and convective drying parameters on kinetics, total phenolic content, carotenoid content and antioxidant capacity of kiwiberry (*Actinidia arguta*). *Appl. Sci.* **2020**, *10*, 6914. [CrossRef]
32. Singleton, V.L.; Rossi, J.A. Colorimetry of Total Phenolics with Phosphomolybdic-Phosphotungstic Acid Reagents. *Am. J. Enol. Vitic.* **1965**, *16*, 144–158. [CrossRef]
33. Dadan, M.; Rybak, K.; Wiktor, A.; Nowacka, M.; Zubernik, J.; Witrowa-Rajchert, D. Selected chemical composition changes in microwave-convective dried parsley leaves affected by ultrasound and steaming pre-treatments—An optimization approach. *Food Chem.* **2018**, *239*, 242–251. [CrossRef] [PubMed]
34. Sledz, M.; Nowak, P.; Witrowa-Rajchert, D. Drying of parsley leaves pre-treated by ultrasound. *Adv. Agric. Sci. Probl. Issues* **2014**, *579*, 91–99.
35. Re, R.; Pellegrini, N.; Proteggente, A.; Pannala, A.; Yang, M.; Rice-Evans, C. Antioxidant activity applying an improved ABTS radical cation decolorization assay. *Free Radic. Biol. Med.* **1999**, *26*(9–10), 1231–1237. [CrossRef]
36. Brand-Williams, W.; Cuvelier, M.E.; Berset, C.L.W.T. Use of a free radical method to evaluate antioxidant activity. *LWT-Food Sci. Technology* **1995**, *28*, 25–30. [CrossRef]
37. Mohapatra, D.; Giri, S.K.; Prasad, S.; Kar, A.; Nema, P.K. Vacuum-microwave drying characteristics of spearmint leaves. *J. Food Res. Technol.* **2014**, *2*, 87–92.
38. Torki-Harchegani, M.; Ghanbarian, D.; Pirbalouti, A.G.; Sadeghi, M. Dehydration behaviour, mathematical modelling, energy efficiency and essential oil yield of peppermint leaves undergoing microwave and hot air treatments. *Renew. Sustain. Energy Rev.* **2016**, *58*, 407–418. [CrossRef]
39. Evin, D. Microwave drying and moisture diffusivity of white mulberry: Experimental and mathematical modeling. *J. Mech. Sci. Technol.* **2011**, *25*, 2711–2718. [CrossRef]
40. Mohammadi, H.; Esmailpour, M.; Gheranpaye, A. Effects of TiO_2 nanoparticles and water-deficit stress on morpho-physiological characteristics of dragonhead (*Dracocephalum moldavica* L.) plants. *Acta Agric. Slov.* **2016**, *107*, 385–396. [CrossRef]
41. Yousefzadeh, S.; Daryai, F.; Mokhtassi-Bidgoli, A.; Hazrati, S.; Yousefzadeh, T.; Mohammadi, K. Morphological, essential oil and biochemical variation of *Dracocephalum moldavica* L. populations. *J. Appl. Res. Med. Aromat. Plants* **2018**, *10*, 59–66. [CrossRef]
42. Sledz, M.; Witrowa-Rajchert, D. Influence of microwave-convective drying of chlorophyll content and colour of herbs. *Acta Agrophysica* **2012**, *19*, 865–876.
43. Di Cesare, L.F.; Forni, E.; Viscardi, D.; Nani, R.C. Changes in the chemical composition of basil caused by different drying procedures. *J. Agric. Food Chem.* **2003**, *51*, 3575–3581. [CrossRef] [PubMed]
44. Di Cesare, L.F.; Forni, E.; Viscardi, D.; Nani, R.C. Influence of drying techniques on the volatile phenolic compounds, chlorophyll and colour of oregano (*Origanum vulgare* L. ssp. *Prismaticum gaudin*). *Ital. J. Food Sci.* **2004**, *16*, 165–175.
45. McSweeney, M.; Seetharaman, K. State of Polyphenols in the Drying Process of Fruits and Vegetables. *Crit. Rev. Food Sci. Nutr.* **2014**, *55*, 660–669. [CrossRef] [PubMed]
46. Chan, E.W.C.; Lim, Y.Y.; Wong, S.K.; Lim, K.K.; Tan, S.P.; Lianto, F.S.; Yong, M.Y. Effects of different drying methods on the antioxidant properties of leaves and tea of ginger species. *Food Chem.* **2009**, *113*, 166–172. [CrossRef]
47. Aprotosoaie, A.C.; Mihai, C.T.; Vochita, G.; Rotinberg, P.; Trifan, A.; Luca, S.V.; Petreus, T.; Gille, E.; Miron, A. Antigenotoxic and antioxidant activities of a polyphenolic extract from European *Dracocephalum moldavica* L. *Ind. Crops Prod.* **2016**, *79*, 248–257. [CrossRef]
48. Nowacka, M.; Śledź, M.; Wiktor, A.; Witrowa-Rajchert, D. Changes of Radical Scavenging Activity and Polyphenols Content During Storage of Dried Apples. *Int. J. Food Prop.* **2014**, *17*, 1317–1331. [CrossRef]
49. Vadivambal, R.; Jayas, D. Changes in quality of microwave treated agricultural products—A review. *Biosyst. Eng.* **2007**, *98*, 1–16. [CrossRef]

applied sciences

MDPI

Article

Microwave and Ultrasound Assisted Rotary Drying of Carrot: Analysis of Process Kinetics and Energy Intensity

Dominik Mierzwa and Grzegorz Musielak *

Division of Process Engineering, Institute of Chemical Technology and Engineering, Poznan University of Technology, ul. Berdychowo 4, 60-965 Poznan, Poland; dominik.mierzwa@put.poznan.pl
* Correspondence: grzegorz.musielak@put.poznan.pl; Tel.: +48-61-665-3622

Abstract: Convective drying is one of the most commonly employed preservation techniques for food. However, the use of high temperatures and extended drying times often leads to a reduction in product quality and increased energy consumption. To address these issues, hybrid processes combining convective drying with more efficient methods are frequently employed. This study investigates the convective rotary drying of carrot (cv. Nantes), assisted by microwaves and ultrasound, using a hybrid rotary dryer. In total, four distinct drying programs—comprising one convective and three hybrid approaches—were evaluated. The study assessed drying kinetics, energy consumption, and product quality. The use of ultrasound increased the drying rate by 13%, microwaves by 112%, and microwaves and ultrasound together by 140%. The use of microwaves reduced energy consumption by 30%, whereas ultrasound resulted in a slight increase. All processes resulted in a significant reduction in water activity. Ultrasound decreased the color difference index, while microwaves increased it compared to convective drying.

Keywords: color difference; drying constant; drying curves; drying rate; drying time; effective diffusion coefficient; mass transfer coefficient; specific energy consumption; temperature; water activity

1. Introduction

One of the world's greatest challenges is the preservation of food. Drying is among the most common methods for preserving fruits and vegetables, effectively reducing moisture content and water activity—key factors in limiting natural decomposition processes. Drying foods limits bacterial and fungal decomposition, resulting in product stabilization [1]. However, most drying techniques involve prolonged heat application, which often compromises product quality and value. Factors like drying method, temperature, humidity, air flow rate, and additional heat sources such as microwave (MW), infrared (IF), or ultrasound (US) influence the extent of quality changes. Convective drying, a widely used method involving hot air, often requires high temperatures and extended durations, which trigger chemical and biochemical reactions that alter color, flavor, aroma, and nutritional properties [2–5]. Drying can also cause excessive shrinkage and/or shape deformation [6–8], making the product less appealing to consumers.

Another significant issue with drying is its high energy consumption. The energy needed to provide sufficient heat for evaporation makes drying one of the most energy-intensive unit operations. It is estimated that drying processes account for up to 25% of total industrial energy consumption [9,10]. Given the environmental and energy concerns, such as fossil fuel depletion and the need to reduce greenhouse gas emissions [11], reducing energy consumption in all industrial sectors, including those utilizing drying processes, is paramount. Therefore, ensuring good product quality while minimizing energy consumption remains a key challenge in drying operations.

One possibility for improving drying processes is the use of pretreatment. Examples of pretreatment methods include osmosis [11–14], ultrasound [15,16], or pulsed electric field [12,17]. Another way to enhance drying processes is to use hybrid drying, which combines two or more techniques in a single process [18]. To accelerate the process, several methods are used in conjunction to supply additional energy to the process during drying. Such acceleration can result in reduced energy consumption and improved product quality. Plant materials are often dried using convection with the addition of microwaves [15,19–21], infrared [15,22,23], or ultrasound [20,24–26]. In most cases, the materials being dried in these processes are stationary, with exceptions being rotary convection–microwave and convection–infrared drying [15,19,22].

The aim of this work was to investigate the effect of the independent use of microwaves and ultrasound, as well as the simultaneous use of both energy sources, on rotary convective drying. Carrot root was chosen as the research material due to its high degree of hardness and inhomogeneous structure, making it particularly challenging to dry [27]. In total, four drying schedules were experimentally tested to determine the most effective process: convective–hot air drying (CV), convective drying assisted by ultrasound (CVUS), convective drying assisted by microwaves (CVMW), and convective drying assisted by both microwaves and ultrasound (CVMWUS). Each schedule was assessed in terms of drying kinetics, product quality, and energy consumption. The use of ultrasound increased the drying rate by 13%, microwaves by 112%, and both sources combined by as much as 140%. All processes resulted in a significant reduction in water activity, with the greatest reduction observed in the CVUS process. Ultrasound also led to a decrease in the color difference index (*dE*), while microwaves significantly increased the value of this index. Additionally, the use of microwaves significantly reduced energy consumption, while ultrasound resulted in a slight increase in energy consumption.

2. Methodology of the Study

2.1. Material

Fresh carrots (*Daucus carota* cv. Nantes), bought at a local market, were used as the experimental material. Before processing, the material was stored in a refrigerator at 4 °C for at least 24 h. Next, three or four of the roots were washed, peeled, and cut into slices (5 mm thick) with the use of an industrial cutter, Hällde RG100 (Kista, Sweden). The prepared samples were divided into two parts. One part was used to measure moisture content, water activity, and color; the other was subjected to drying. The prepared batch of material, weighing approximately 200 g, was placed into the drum of the dryer and subjected to the drying process. The dryer was not preheated prior to the beginning of the drying process.

The initial moisture content of the material was determined using a Precisa XM120 moisture analyzer (Dietikon, Switzerland) and equaled 86.63 ± 1.73% (wet basis).

2.2. Drying Procedure

The drying tests were conducted using a laboratory rotary hybrid dryer constructed by PROMIS-TECH (Wrocław, Poland), as shown schematically in Figure 1.

The dryer has two different techniques for drying: hot air-convective (CV) and microwave (MW). Convective drying utilizes air drawn in by a blower (Figure 1(1)) from the dryer's environment, which, after passing through a heater (5), is forced directly into the dryer drum (12). After passing through the drum, the moistened air leaves the dryer through the outlet duct (7b). Microwave drying is implemented using electromagnetic waves at 2.45 GHz, generated continuously by two water-cooled magnetrons (10), produced by Muegge (Reichelsheim, Germany). Additionally, each of the drying techniques may be supported by high-power ultrasound (US) generated by a transducer mounted on the side of the dryer (13) and delivered through the air directly to the material. In the experiment, the frequency of the acoustic waves was 26 kHz, and the maximum acoustic intensity reached 160 dB. The operation of the transducer is controlled by two of the airborne ultrasound system's (AUS) other units, specifically the controller (2) and the preamplifier (3). The system was developed and manufactured by Pusonics (Madrid, Spain).

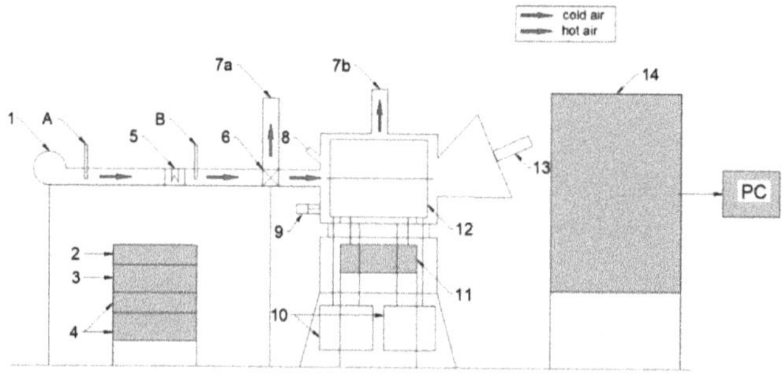

Figure 1. Scheme of the rotary hybrid dryer: 1—blower, 2—Airborne Ultrasound system (AUS) controller, 3—AUS amplifier, 4—microwave feeders, 5—heater, 6—pneumatic valve, 7a and 7b—air outlet, 8—pyrometer, 9—drum drive, 10—microwave generators, 11—balance, 12—rotating drum, 13—AUS transducer, 14—control unit, A—temperature and humidity sensor, B—temperature sensor.

The research included four drying schemes. Each drying schedule was elaborated on the basis of the convective process (CV) supported by microwaves (MW) and/or ultrasound (US). The temperature and the velocity of the drying agent (air) was constant during the whole research process. The power of the microwave and ultrasonic enhancement was selected in accordance with the results of preliminary experiments. Each drying procedure was carried out in triplicate. Details about the drying conditions for the specified schedules of drying are given in Table 1.

Table 1. Drying procedures.

No.	Acronym	Air Parameters			Radiation Power (W)	
		T (°C)	v (m/s)	RH (%)	US	MW
1	CV				0	0
2	CVUS	60	0.2	7	200	0
3	CVMW				0	100
4	CVMWUS				200	100

T—air temperature, v—air velocity, RH—air relative humidity, US—ultrasound, MW—microwave.

The weight loss of the samples being dried was measured automatically throughout the whole operation at 5 min intervals using a laboratory balance (Figure 1(11)) model APP/25c (precision 0.01 g), produced by Radwag (Radom, Poland). To prevent any potential interference that might affect the accuracy of the mass measurement, the drum of the dryer was temporarily stopped, and the drying air was redirected through a pneumatic valve (6) to the outlet (7a), thus bypassing the chamber. The drying agent's parameters, such as temperature, velocity, and humidity, were also measured throughout the process using two sensors (A, B) models HD29371TC1.5 and HD4817ETC1.5 (precision 0.01 m/s, 0.1 °C, 0.01%), produced by DeltaOHM (Selvazzano Dentro, Italy). All measured data were collected by data acquisition software running on a personal computer (PC).

The temperature of the samples during drying was measured using two methods, depending on the process variant. In the convective and convective-ultrasound (CV, CVUS) processes, the temperature (*T*) was measured using a standalone HTDL-20 recorder produced by DWYER (Michigan City, IN, USA). Before the test, a single slice of carrot was impaled on the rigid external probe of the device and the whole was placed in the dryer with the other samples. After the drying process, the temperature recorded over time (at one-minute intervals) was read from the device. For the convective-microwave processes (CVMW, CVMWUS), the surface temperature (*T**) was measured using a model B1 infrared thermal camera made by FLIR (Wilsonville, OR, USA). At preset intervals (every 10 min), the drying process was paused in order to take a thermal image—a thermogram—of the samples inside the dryer. The images were further analyzed using software provided by the camera manufacturer. The point with the highest temperature was marked on the images and its value was read.

2.3. Drying Kinetics and Energy Intensity

Drying kinetics were evaluated on the basis of drying curves, that is, changes in dimensionless moisture content over time, and changes in drying rate as a function of dimensionless moisture content. In addition, the average drying rate was also determined for each process.

By defining the dry basis moisture content as the ratio of the mass of moisture (m_w) contained in the material to the mass of the dry phase (m_d) [28]:

$$X = m_w/m_d, \tag{1}$$

the drying kinetics of samples could be assessed in terms of dimensionless moisture content (Y) and drying rate (DR) in accordance with the following equations [29]:

$$Y_i = \overline{X}(t_i)/X_0, \tag{2}$$

$$DR_i = -m_d(dX/dt) = -(dm_w/dt) = -(m(t_i) - m(t_{i-1}))/(t_i - t_{i-1}), \tag{3}$$

where $\overline{X}(t_i)$ is the spatial average of the moisture content determined at time t_i, X_0 is the initial moisture content, dm_w is the weight loss of the sample, and dt is the time at which the weight loss dm_w occurred.

The mass of moisture m_w can be defined as the difference between the mass of the moist material m and the mass of the dry phase m_d. After substituting this difference into (1) the spatial average moisture content $\overline{X}(t_i)$ (dry basis) at any i-th time of the process was calculated using the following formula [30]:

$$\overline{X}(t_i) = (m(t_i) - m_d)/m_d, \tag{4}$$

where $m(t_i)$ is the sample mass measured at time t_i, and m_d is the dry matter determined using a Precisa XM120 moisture analyzer (Dietikon, Switzerland) with an accuracy of 0.01%.

The average drying rate for a given drying process was calculated using the following equation [30]:

$$DR_{av} = \Delta m/DT, \tag{5}$$

where Δm is the mass of moisture removed from the sample by drying, and DT is the total drying time.

The experimental data was also used to determine the drying constant, k; mass transfer coefficient, α; and effective diffusion coefficient, D_{eff}. All these parameters could be determined from the drying rate curves [30]. Based on the experimental results, the values of the average moisture content $\left(\overline{X}\right)_i$ and the rate of moisture content change $-\left(d\overline{X}/dt\right)_i$ were calculated in accordance with the following equations [30]:

$$\left(\overline{X}\right)_i = \left(\overline{X}(t_i) + \overline{X}(t_{i-1})\right)/2, \tag{6}$$

$$-\left(d\overline{X}/dt\right)_i = \left(\overline{X}(t_{i-1}) - \overline{X}(t_i)\right)/(t_i - t_{i-1}), \tag{7}$$

The rate of moisture content change $-\left(d\overline{X}/dt\right)_i$ (7) was calculated as an average value for the time period $(t_i - t_{i-1})$. For consistency in kinetics modeling, material moisture content $\left(\overline{X}\right)_i$ (6) was also averaged over the same time period $(t_i - t_{i-1})$. Based on the linear approximation of the moisture content change over average moisture content, the drying constant, k, was determined to be the slope of approximation lines. The mass transfer, α, and the effective diffusion coefficient, D_{eff}, were then calculated using the following formula [30]:

$$\alpha = kl, \tag{8}$$

$$D_{eff} = \left(4l^2k\right)/\pi^2, \tag{9}$$

where l is half of the sample's thickness.

Each drying schedule was also assessed in terms of energy intensity. The total electric energy consumed by the experimental setup during the drying tests was measured using an electricity meter. The resulting electricity consumption, EC, was then transformed through recalculation into specific energy consumption, SEC, that is, the amount of energy used to evaporate one kilogram of moisture according to the relationship:

$$SEC = (3.6 \cdot EC)/\Delta m, \tag{10}$$

The value thus determined enabled the energy consumption of each process to be compared.

2.4. Quality of Products

The quality of the obtained products was evaluated on the basis of water activity and the color of the samples. Water activity (a_w) was measured at room temperature (~23 °C) before and after drying experiments with a TESTO 650 multimeter (Testo, Titisee-Neustadt, Germany), equipped with a water activity measuring chamber. Prior to the measurement, approximately 5 g of ground sample was placed in a polystyrene container and subsequently inserted into a steel airtight measuring chamber. The measurement was conducted until a constant a_w value was reached, indicating no change at a value of 0.001 for a period of five minutes.

The color of the samples was measured using a CR400 colorimeter produced by Konica Minolta (Tokyo, Japan) and quantified in terms of the three-dimensional color space, CIELAB, where L^* represents brightness, a^* denotes the chromatic component of colors ranging from red to green, and b^* signifies the chromatic component of colors ranging from yellow to blue. Based on the measured values for the raw material (index $_0$) and the dried sample (index $_d$), the color difference index, dE, was determined according to the following equation [31]:

$$dE = \sqrt{\left(L_0^* - L_d^*\right)^2 + \left(a_0^* - a_d^*\right)^2 + \left(b_0^* - b_d^*\right)^2} \tag{11}$$

The experimental data were processed using OriginPro 2024b [32]. The data presented in Section 3, Results and Discussion, are means ± standard deviations.

3. Results and Discussion

3.1. The Kinetics and Energy Intensity of Drying

Figure 2a shows a graph demonstrating the dependence of dimensionless moisture content (Y) on time (t). Despite the variations in drying methods, the curve paths are very similar. The characteristic drying periods such as the constant drying-rate period and the falling drying-rate period cannot easily be distinguished, which is typical for biomaterials drying processes [9,33]. As expected, convective drying alone was the slowest among the methods tested. However, the application of microwaves (MW) or ultrasound (US) significantly reduced the drying time by increasing the rate of drying. Figure 2b illustrates the relationship between the drying rate (DR) and dimensionless moisture content (Y) during the process.

A comparative analysis of the drying rate curves reveals that the application of microwaves had a significantly greater impact than ultrasound (Figure 2b). The maximum drying rate for the microwave processes (CVMW and CVMWUS) reached a value above 0.03 g/s, while for the convective–ultrasound process (CVUS), the highest observed value was 0.02 g/s. The lowest drying rate was observed for the convection process (CV), with a maximum of approximately 0.015 g/s, which resulted in a prolonged drying time.

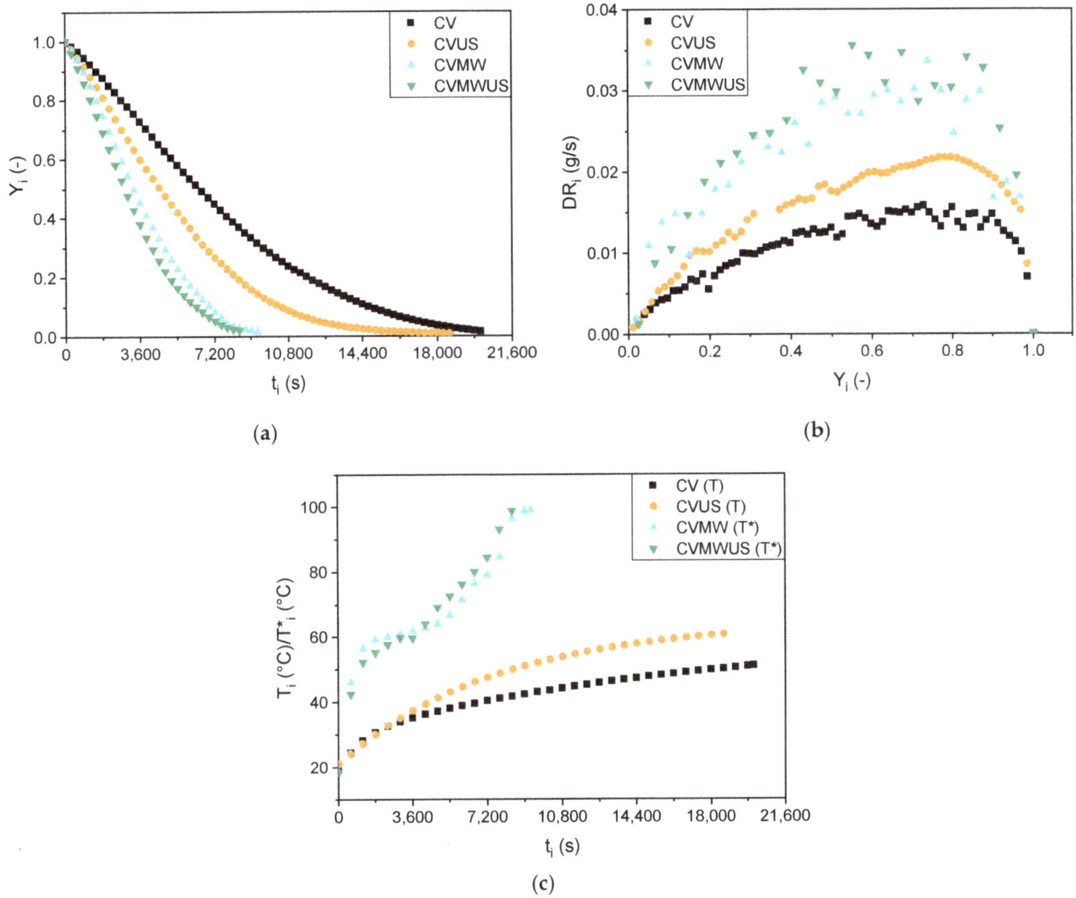

Figure 2. Evolution of dimensionless moisture content, Y_i (**a**); drying rate, DR_i (**b**); and temperature of the samples, T/T^* (**c**), for each process.

The shape of the drying rate curves exhibits a similar pattern across the different processes, which enables the essential drying periods to be identified. These include the initial stage, the constant drying rate period, and the falling drying rate period. The values of Y_i at which the transition between the drying periods occurs are slightly different for each process; however, some similarities can be observed between the CV and CVUS processes, as well as between the CVMW and CVMWUS processes.

The initial stage of the process lasts from the beginning until a constant drying rate is reached. For the CV-CVUS pair of processes, this period lasted until the dimensionless moisture content reached $Y_i \approx 0.89$, while for the CVMW-CVMWUS pair, it lasted until $Y_i \approx 0.87$. In both instances, this occurred after approximately 1200 s of the process' duration. The relatively prolonged heating period observed in this study may be attributed to two factors: the larger sample batch size compared to other studies on hybrid drying and the initiation of drying on the cold dryer. In this approach, a portion of the heat energy supplied with the hot air was utilized for the heating of the dryer components. The process subsequently entered the constant drying rate period. In the case of the CV-CVUS pair of processes, this period lasted until the material reached a dimensionless moisture content of approximately 0.57, which occurred after 5400 s of the convection (CV) process and 3900 s of the ultrasound-assisted process (CVUS). In the microwave-assisted processes

(CVMW, CVMWUS), the constant drying rate period ended at a dimensionless moisture content of 0.47, which was reached after approximately 3600 s, despite the variations in the types of processes. The condition for the end of the period of constant drying rate is the slowing down of the transport of moisture from the inside of the sample to its surface. In all the processes studied, the transport of moisture to the surface of the sample is caused by diffusion. When microwaves are used (CVMW, CVMWUS), the temperature of the samples is higher than in the other processes (CV, CVUS). This facilitates diffusion at lower moisture contents. In addition, internal heating of the material with microwaves causes an increase in moisture pressure, which in turn causes easier transport of moisture to the surface (pressure flow). For these two reasons, the period of constant drying rate is longer in the processes with microwaves (CVMW, CVMWUS) than in the processes without microwaves (CV, CVUS). Once the aforementioned moisture content—frequently designated as the critical moisture content—is surpassed, a falling drying rate period ensues, persisting until the end of the process.

By analyzing the temperature curves (Figure 2c), it can be seen that in the microwave-assisted processes (CVMW, CVMWUS) the temperature of the material increased much faster and reached a significantly higher value than for the convection (CV) or ultrasonic-assisted (CVUS) processes. This phenomenon is due to the effective interaction between the microwaves and the moist material being dried, and results in high drying rates and short processing times. The temperature curves also demonstrate the beneficial impact of ultrasound on heat transfer. The application of acoustic waves in the CVUS process resulted in a significantly enhanced heating rate when compared to the convection process (CV). This phenomenon may be attributed to ultrasound-induced effects at the material's surface, which result in a reduction of the boundary layer's thickness and thus enhances heat transfer efficiency and the absorption of ultrasonic energy [26].

An analysis of the shape of the temperature curves (Figure 2c) also reveals the similarity between the CV and CVUS processes, and between the CVMW and CVMWUS processes. In the case of convective (CV) and ultrasound-assisted convective (CVUS) drying, there are no evident characteristic periods, and the curves move asymptotically towards the temperature of the drying agent. For the microwave processes (CVMW and CVMWUS), three stages can be distinguished. During the first, there was intensive heating to a temperature of about 60 °C. This stage lasted about 1200 s and corresponds to the initial stage on the drying rate curves (Figure 2b). During the next stage, the temperature of the material was approximately constant (about 60 °C) and lasted up to 3600 s; this coincides with the constant drying rate period (Figure 2b). During the last stage, the temperature increased again to reach a value of about 100 °C at the end of the process. This period corresponds to a falling drying rate period.

Concerning the microwave processes, the intense heating during the initial stage was the result of effective absorption of microwave radiation by the very moist material. In this stage, intense moisture evaporation also began, leading to an increase in the drying rate (Figure 2b). The constant temperature of the material during the constant drying rate period (from 1200 s to 3600 s of the process) was due to the very intensive evaporation of moisture from the sample's surface. Because of this, convectively supplied and microwave-generated heat is used to evaporate moisture and does not heat up the material being dried. This phenomenon occurs only when there is enough moisture in the material, so that it can be transported efficiently to the surface/evaporation front, and that transport within the material does not limit the process rate. Given the constant value of the heat of vaporization, the rate of drying (evaporation) is thus primarily determined by the amount of energy applied per unit time. If drying is conducted under constant process conditions (i.e., constant air temperature and radiation power), it can be assumed that the amount of energy supplied to the surface of the material being dried is also constant. Consequently, the evaporation rate remains constant, resulting in a constant drying rate (Figure 2b) and also in almost constant temperature of material [34]. Once the moisture content of the material has decreased below a critical value, the effectiveness of the moisture transport to the surface/evaporation front drops. The energy applied to the material is then used not only to evaporate the moisture but also to heat the material, resulting in a temperature increase (Figure 2c). As the drying rate during this period is determined primarily by the internal transport of moisture, the value of this parameter continues to decrease (Figure 2b) and falling drying rate period occurs.

The notable occurrence of a constant drying rate period may be somewhat unexpected. A review of the literature reveals numerous reports concerning materials of plant origin for which a constant drying rate period is not observed, and the entire drying process occurs during the period of falling drying rate [9]. However, some researchers have reported the presence of a constant drying rate period. Arslan et al. reported similar drying curves for carrot [35] and red pepper [36]. On the other hand, Abbaspour-Gilandeh et al. [15] did not observe a constant drying rate period during the drying of carrot using multiple techniques. The underlying reason behind this discrepancy remains unclear. May and Perré [33] hypothesized that this phenomenon may be linked to a variation in the mass exchange area resulting from material shrinkage during the drying process. However, determining exchange surface area is extremely difficult and requires advanced experimental techniques.

In Figure 3, the average drying rate, DR_{av}; time, DT_{av}; and specific energy consumption, SEC_{av}, for each particular drying process are presented.

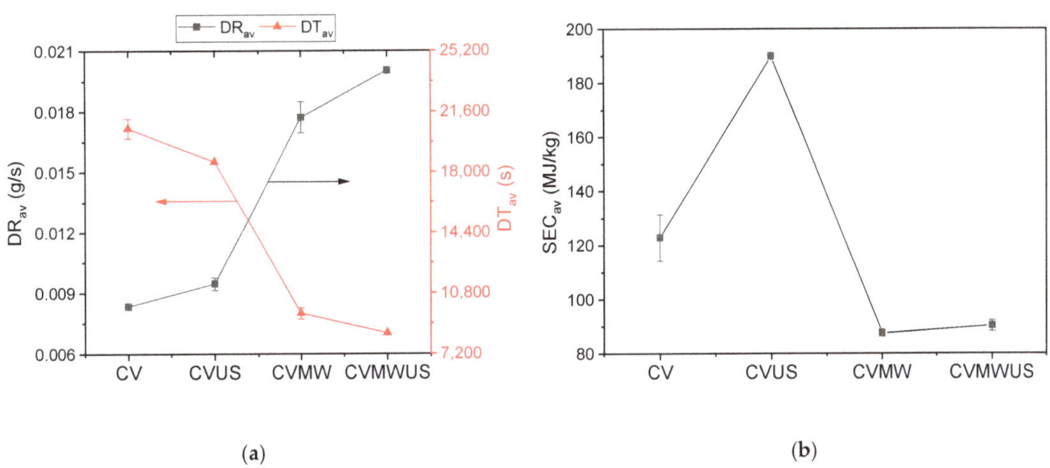

Figure 3. Average drying rate, DR_{av} and time, DT_{av} (**a**); and specific energy consumption, SEC_{av} (**b**) for each process.

The average drying rate obtained during the hybrid processes was visibly higher compared to the convective process (Figure 3a). In the case of the fastest drying procedure (CVMWUS) the average drying rate increased by 140% compared to CV, while for other hybrid programs, namely ultrasound–assisted (CVUS) and microwave–assisted (CVMW) convective drying, the increase equaled 13% and 112%, respectively. These changes were reflected in the value of the average drying time, which exhibited reductions of 10%, 53%, and 59% for CVUS, CVMW, and CVMWUS, respectively (Table 2). A strong positive correlation (Pearson's $r = 0.99$) was identified between the DR_{av} and DT_{av} values; a finding that is also evident in the graphical representation (Figure 3a).

Table 2. Relative change in average drying rates and specific energy consumption.

Process	Relative Change		
	DR_{av}	DT_{av}	SEC_{av}
CV	–	–	–
CVUS	13%	−10%	59%
CVMW	112%	−53%	−30%
CVMWUS	140%	−59%	−28%

On the basis of the obtained results, microwaves were found to be a more efficient enhancement agent compared to ultrasound; this follows from the different mechanisms of interaction they have with the material being dried. Microwaves heat the entire volume of the sample, thereby significantly intensifying the transport of moisture within the body (through the intensification of diffusion and evaporation). Furthermore, the application of MW radiation typically results in a reduction in the temperature gradient within the material, which additionally enhances the drying process (through the reduction of thermodiffusion) [34]. The obtained results are in good agreement with the reports of other authors. Abbaspour-Gilandeh et al. [15] also observed a 50% reduction in drying time for microwave-assisted hot air drying of carrot. Kroehnke et al. [37] noted a fivefold decrease in drying time during the microwave-assisted convective processing of this vegetable.

The influence of ultrasound on the material being dried is very limited and depends strictly on the acoustic impedance mismatch between the sample material and the surrounding air. Several authors have proposed different mechanisms for ultrasound interaction, for example, cavitation, the "sponge effect", microstreaming, the formation of microchannels, disturbance in the boundary layer, etc. [26]. The proposed phenomena may accelerate the drying operation by reducing both the internal and external resistances. On the basis of the results obtained during the authors' previous research [38,39], it can be stated that in the case of the presented drying system, ultrasound influences the kinetics of the process rather by reducing external resistance factors such as the laminar boundary layer. Due to the significant losses in ultrasound power/intensity that occur during the transfer of energy between the radiator plate and the material, only a small proportion of the acoustic energy is involved in accelerating the drying process. Nevertheless, the positive effect of the ultrasound assistance could be easily distinguished and is in agreement with the results presented by other researchers. Kroehnke et al. [37] noted a 20% reduction in processing time during the ultrasound-assisted convective drying of carrot.

An analysis of the specific energy consumption (*SEC*) (Figure 3b) reveals that the application of additional energy-consuming devices, in the form of microwave generators or ultrasound transducer, affects the energy intensity of the process in a different manner. In the case of the microwave processes (CVMW, CVMWUS), a notable enhancement of the drying rate—which also resulted in a considerable reduction in drying time—led to a reduction in the amount of energy required to evaporate a given mass of moisture. The SEC_{av} values for these processes were the lowest for all the schemes studied: 87 MJ/kg and 90 MJ/kg for CVMW and CVMWUS, respectively. This represents a respective 30% and 28% reduction compared to the convection CV process (Table 2). The ultrasound-assisted convective process (CVUS), despite an observable increase in drying rate and reduction in drying time, did not prove to be more energy efficient than the reference process. A notable increase (59%) in the value of specific energy consumption was observed for CVUS, reaching an approximate value of 190 MJ/kg.

The considerable increase in *SEC* for CVUS can be attributed to the higher instantaneous energy consumption (in comparison to CV) necessitated by the requirement to power the AUS ultrasound generation system. Despite the beneficial effects of acoustic waves on the drying rate and time, their effectiveness was insufficient to compensate for additional energy expenditure, leading to an increase in total energy consumption. This phenomenon was previously reported by Kowalski et al. [38], Kroehnke et al. [37], and Kroehnke and Musielak [40].

In the case of the CVMW and CVMWUS processes, the instantaneous energy consumption was also higher. However, the kinetic benefits resulting from the use of microwave radiation (increased drying rate and reduced drying time) were substantial enough for compensate this additional energy expenditure and reduce overall energy consumption. The discrepancy between the *SEC* values for the ultrasound- and microwave-assisted processes may also be attributed to the differing energy efficiency of the generators and the varying attenuation coefficient of the radiation along the path from the generator/transducer. The energy efficiency of microwave-assisted processes was also confirmed in studies by Abbaspour-Gilandeh et al. [15] and Kroehnke et al. [37]. In both cases, the microwave-assisted drying of carrot was characterized by a lower energy consumption compared to the reference convective drying.

The experimental data was also used to determine the drying constant, k; mass transfer coefficient, α; and effective diffusion coefficient, D_{eff}. Table 3 presents the values obtained for each process.

Table 3. Values for drying constant, k; mass transfer, α; and effective diffusion coefficient, D_{eff}.

Process	k (1/s)	α (m/s)	D_{eff} (m²/s)	adj. R^2 (–)
CV	$1.75 \times 10^{-4} \pm 6.02 \times 10^{-6}$	$4.39 \times 10^{-7} \pm 1.50 \times 10^{-8}$	$4.44 \times 10^{-10} \pm 1.52 \times 10^{-11}$	0.9612
CVUS	$2.17 \times 10^{-4} \pm 5.50 \times 10^{-6}$	$5.43 \times 10^{-7} \pm 1.38 \times 10^{-8}$	$5.50 \times 10^{-10} \pm 1.39 \times 10^{-11}$	0.9619
CVMW	$3.29 \times 10^{-4} \pm 2.07 \times 10^{-5}$	$8.23 \times 10^{-7} \pm 5.17 \times 10^{-8}$	$8.34 \times 10^{-10} \pm 5.23 \times 10^{-11}$	0.9243
CVMWUS	$4.59 \times 10^{-4} \pm 2.79 \times 10^{-5}$	$1.15 \times 10^{-6} \pm 6.99 \times 10^{-8}$	$1.16 \times 10^{-9} \pm 7.08 \times 10^{-11}$	0.9520

The values for all determined parameters corroborate the hypothesis that ultrasonic and microwave assistance has a favorable impact on convective drying kinetics. For the hybrid processes, the drying constant, mass transfer, and effective diffusion coefficient are markedly higher than for the convective process. There is also a clear distinction between acoustic waves and microwave radiation. In the case of the latter, the observed values are higher than in the case of ultrasound, which confirms the fact that it has a more effective influence on mass transfer processes. Arslan et al. [35] and Abbaspour-Gilandeh et al. [15] also observed significant increases in the effective diffusion coefficient for the microwave drying of carrot. On the other hand, Kroehnke et al. [37] noted an increase in the convective mass transfer coefficient with an increase in the power of the ultrasound enhancement. It can therefore be stated that the intensification of mass exchange is less pronounced for ultrasound than for microwaves, yet it is discernible and has a beneficial impact on convective drying kinetics. In turn, the values of the adjusted coefficient of determination (adj. R^2) confirm that there is close convergence between the experimental curve and the linear approximation (Table 3).

3.2. Assessment of Product Quality

The water activity (a_w) of a food product is a crucial factor in determining its shelf life. A low a_w inhibits the growth of pathogenic microorganisms, such as fungi, yeasts, and molds, and prevents unfavorable biochemical reactions, including enzymatic and non–enzymatic processes, and oxidation. It is generally accepted that microbial activity ceases below an a_w value of 0.6, while the most unfavorable biochemical reactions are maximally inhibited for an a_w below 0.4 [41]. Figure 4 depicts the mean water activity values for each dried sample and the raw material.

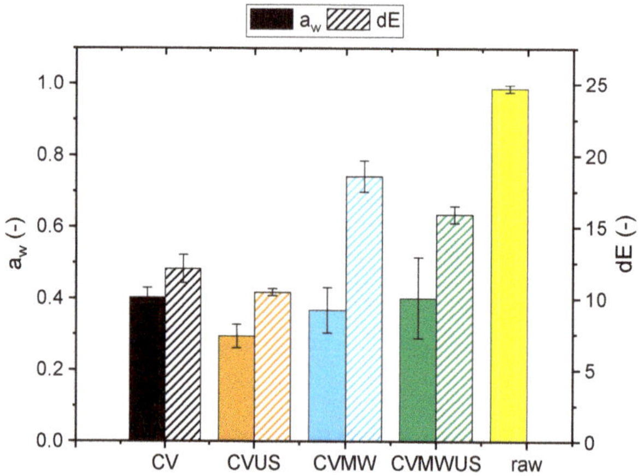

Figure 4. Average water activity, a_w, and color difference index, dE, for each process.

From an analysis of the results, we can conclude that the safety and microbiological stability of the material were ensured, as the results for the obtained (dried) products were characterized by a water activity (a_w) value below 0.4. Differences between the particular drying procedures were negligible and did not affect the shelf-life of the material.

The color of the samples is also a significant quality parameter. It is primarily responsible for determining the sensory properties of the product and indicating the so-called palatability. Additionally, color can indirectly indicate the safety of a product or the changes that occur in it, such as the color change in fruit during the ripening process. It must be acknowledged, however, that color perception is highly subjective, and the evaluation of products will invariably depend on the observer's preferences. Consequently, instrumental color measurement was introduced as a means of objectively determining changes in color relative to the raw material. Figure 4 illustrates the color differences (dE) for dried samples in relation to the raw material. It is assumed that a value above 4 units represents a perceptible difference in color as discerned by an average observer.

As can be observed in Figure 4, all the dried products exhibited a color that was discernibly distinct from that of the raw material, as indicated by color difference index values exceeding 4 units. The application of microwaves in microwave-assisted convective drying (CVMW and CVMWUS) resulted in the most detrimental impact on product quality. The color difference index (dE) was, for these processes, visibly greater in comparison to the solely convective (CV) or ultrasound-assisted convective processes (CVUS) (Figure 4). Thus it turned out that the very significant achievements made in the drying kinetics led to the opposite results regarding the quality of the products. This is often observed in drying processes, for example, as found by Mierzwa and Szadzińska [39].

The color loss that occurs during microwave drying is usually caused by the high temperatures to which the products are exposed. Figure 5 shows thermal images specifically of samples taken at the 140th minute of the CVMW and CVMWUS drying processes. The maximum material temperature recorded in the image is approximately 98 °C, indicating a considerable degree of overheating in the material. Such elevated temperatures give rise to a number of detrimental changes, which, among other parameters, impact the color. These changes are attributed to both biochemical processes and thermal decomposition [42]. The rapid increase in temperature of the material being microwave-dried can be attributed to the inhomogeneity of the microwave field and changes occurring in the material. In the former, localized overheating of the material, known as "hot spots", results in substantial qualitative alterations at these locations. To address this phenomenon, advanced microwave chamber designs are employed, and the dried material's movement is maintained through the microwave field (transmission belt, rotating pan, drum, etc.) [43].

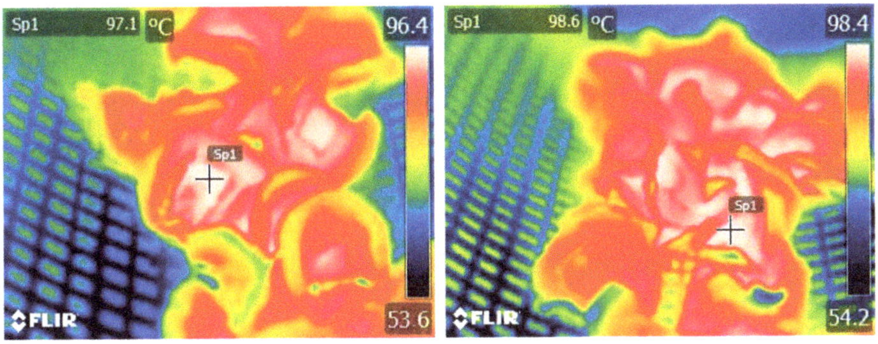

(**a**) (**b**)

Figure 5. Exemplary thermograms of samples dried in microwave–assisted processes: (**a**) CVMW and (**b**) CVMWUS at 140th minute of the processes.

Additionally, changes in moisture content can also result in the material overheating. This is due to the fact that the moisture affects the dielectric properties of the material, namely, its capacity to absorb and reflect microwaves. When the material contains a substantial amount of moisture and the radiation power is properly adjusted, microwaves can effectively evaporate moisture without causing a sudden increase in the material's temperature. However, as the moisture content decreases during the drying process, the ability of the material to absorb microwaves changes. This can lead to a significant increase in the temperature of the material and unfavorable changes in its color. Therefore, it is very important to regulate the microwave power and exposure time, especially towards the end of the process. An alternative solution is the periodic application of microwaves (intermittent microwave drying) and the introduction of a so-called "relaxation period", during which the temperature and humidity gradients will be equalized. This non-stationary hybrid drying is a promising technology that allows operational time to be reduced while maintaining high product quality, including color [35,39,44].

The beneficial impact of ultrasound on product color (Figure 4, CVUS) can be attributed to its non-thermal nature. Although ultrasound application accelerated heating of the material (compared to CV), unlike microwave radiation, it did not result in overheating effects (Figure 2c). Consequently, the initial color of the samples was preserved more effectively, and the majority of observed changes can be attributed to the temperature of the drying agent (air) and the high oxygen content in the atmosphere. These findings align well with those reported by Kroehnke et al. [37].

4. Conclusions

The application of microwaves and/or ultrasound during the convective drying of carrots positively influenced the drying kinetics. The hybrid processes exhibited higher drying rates, resulting in shorter drying times. Among these methods, microwaves were more effective than ultrasound across all evaluated aspects, including drying rate, time, effective diffusion coefficient, and drying constant.

Microwave-assisted drying significantly improved energy efficiency by reducing drying time, leading to lower energy consumption per unit mass of moisture evaporated. In contrast, ultrasound-assisted drying showed higher specific energy consumption compared to conventional convective drying, attributed to its lower drying intensification efficiency and higher instantaneous energy use.

A negative impact of constant microwave irradiation on the color of the product was also observed. In contrast, non-thermal ultrasound waves were found to have minimal impact on the color of the material, thus acting as a quality promoter. All dried products exhibited low water activity, ensuring their safety and microbiological stability. Further research is necessary to investigate the influence of MW and US enhancement as a function of applied power.

Author Contributions: Conceptualization, D.M.; methodology, D.M. and G.M.; formal analysis, D.M.; investigation, D.M.; writing—original draft preparation, D.M. and G.M.; writing—review and editing, D.M. and G.M.; visualization, D.M. All authors have read and agreed to the published version of the manuscript.

Funding: This research was financially supported by the Ministry of Science and Higher Education in Poland.

Institutional Review Board Statement: Not applicable.

Informed Consent Statement: Not applicable.

Data Availability Statement: The raw data supporting the conclusions of this article will be made available by the authors on request.

Conflicts of Interest: The authors declare no conflicts of interest. The funders had no role in the design of the study; in the collection, analyses, or interpretation of data; in the writing of the manuscript; or in the decision to publish the results.

References

1. Jayaraman, K.S.; Das Gupta, D.K. Drying of Fruits and Vegetables. In *Handbook of Industrial Drying*; Mujumdar, A.S., Ed.; CRC Press: Boca Raton, FL, USA, 2014; pp. 611–636, ISBN 978-1-4665-9665-8.
2. Bonazzi, C.; Dumoulin, E. Quality Changes in Food Materials as Influenced by Drying Processes. In *Modern Drying Technology, Volume 3: Product Quality and Formulation*; Wiley-VCH Verlag GmbH & Co. KGaA: Weinheim, Germany, 2011; Volume 3, pp. 1–20, ISBN 978-3-527-63166-7.
3. Krokida, M.K.; Tsami, E.; Maroulis, Z.B. Kinetics on Color Changes During Drying of Some Fruits and Vegetables. *Dry. Technol.* **1998**, *16*, 667–685. [CrossRef]
4. Markowski, M.; Stankiewicz, I.; Zapotoczny, P.; Borowska, J. Effect of Variety on Drying Characteristics and Selected Quality Attributes of Dried Carrots. *Dry. Technol.* **2006**, *24*, 1011–1018. [CrossRef]
5. Santos, P.H.S.; Silva, M.A. Retention of Vitamin C in Drying Processes of Fruits and Vegetables—A Review. *Dry. Technol.* **2008**, *26*, 1421–1437. [CrossRef]
6. Ramos, I.N.; Brandão, T.R.S.; Silva, C.L.M. Structural Changes During Air Drying of Fruits and Vegetables. *Food Sci. Technol. Int.* **2003**, *9*, 201–206. [CrossRef]
7. Ratti, C. Shrinkage during Drying of Foodstuffs. *J. Food Eng.* **1994**, *23*, 91–105. [CrossRef]
8. Zogzas, N.P.; Maroulis, Z.B.; Marinos-Kouris, D. Densities, Shrinkage and Porosity of Some Vegetables During Air Drying. *Dry. Technol.* **1994**, *12*, 1653–1666. [CrossRef]
9. Jangam, S.V.; Mujumdar, A.S. Basic Concepts and Definitions. In *Drying of Foods, Vegetables and Fruits*; Jangam, S.V., Law, C.L., Mujumdar, A.S., Eds.; Singapore, 2010; Volume 1, pp. 1–30, ISBN 978-981-08-6759-1. Available online: https://arunmujumdar.com/wp-content/uploads/2020/03/Drying-of-Foods-Vegetables-and-Fruits-Volume-1.pdf (accessed on 15 November 2024).
10. Strumiło, C.; Jones, P.L.; Żyłła, R. Energy Aspects in Drying. In *Handbook of Industrial Drying*; Mujumdar, A.S., Ed.; CRC Press: Boca Raton, FL, USA, 2014; pp. 1077–1100, ISBN 978-1-4665-9665-8.
11. Grabowski, S.; Marcotte, M.; Poirier, M.; Kudra, T. Drying Characteristics of Osmotically Pretreated Cranberries—Energy and Quality Aspects. *Dry. Technol.* **2002**, *20*, 1989–2004. [CrossRef]
12. Amami, E.; Khezami, L.; Vorobiev, E.; Kechaou, N. Effect of Pulsed Electric Field and Osmotic Dehydration Pretreatment on the Convective Drying of Carrot Tissue. *Dry. Technol.* **2008**, *26*, 231–238. [CrossRef]
13. Lenart, A.; Lewicki, A. Kinetics of Osmotic Dehydration of Plant Tissue. In *Drying '87*; Mujumdar, A.S., Ed.; Hemisphere Publisher: New York, NY, USA, 1987; pp. 239–248.
14. Wang, R.; Zhang, M.; Mujumdar, A.S. Effect of Osmotic Dehydration on Microwave Freeze-Drying Characteristics and Quality of Potato Chips. *Dry. Technol.* **2010**, *28*, 798–806. [CrossRef]
15. Abbaspour-Gilandeh, Y.; Kaveh, M.; Aziz, M. Ultrasonic-Microwave and Infrared Assisted Convective Drying of Carrot: Drying Kinetic, Quality and Energy Consumption. *Appl. Sci.* **2020**, *10*, 6309. [CrossRef]
16. Rawson, A.; Tiwari, B.K.; Tuohy, M.G.; O'Donnell, C.P.; Brunton, N. Effect of Ultrasound and Blanching Pretreatments on Polyacetylene and Carotenoid Content of Hot Air and Freeze Dried Carrot Discs. *Ultrason. Sonochem.* **2011**, *18*, 1172–1179. [CrossRef] [PubMed]
17. Ade-Omowaye, B.I.O.; Angersbach, A.; Taiwo, K.A.; Knorr, D. Use of Pulsed Electric Field Pre-Treatment to Improve Dehydration Characteristics of Plant Based Foods. *Trends Food Sci. Technol.* **2001**, *12*, 285–295. [CrossRef]
18. Kudra, T.; Mujumdar, A.S. Classification and Selection Criteria: Conventional versus Novel Technologies. In *Advanced Drying Technologies*; CRC Press: Boca Raton, FL, USA, 2009; ISBN 978-0-429-15075-3.

19. Kaveh, M.; Abbaspour-Gilandeh, Y.; Nowacka, M. Optimisation of Microwave-Rotary Drying Process and Quality Parameters of Terebinth. *Biosyst. Eng.* **2021**, *208*, 113–130. [CrossRef]
20. Szadzińska, J.; Łechtańska, J.; Kowalski, S.J.; Stasiak, M. The Effect of High Power Airborne Ultrasound and Microwaves on Convective Drying Effectiveness and Quality of Green Pepper. *Ultrason. Sonochem.* **2017**, *34*, 531–539. [CrossRef] [PubMed]
21. Zhang, M.; Jiang, H.; Lim, R.X. Recent Developments in Microwave-Assisted Drying of Vegetables, Fruits, and Aquatic Products—Drying Kinetics and Quality Considerations. *Dry. Technol.* **2010**, *28*, 1307–1316. [CrossRef]
22. Kaveh, M.; Abbaspour-Gilandeh, Y. Impacts of Hybrid (Convective-Infrared-Rotary Drum) Drying on the Quality Attributes of Green Pea. *J. Food Process Eng.* **2020**, *43*, e13424. [CrossRef]
23. Łechtańska, J.M.; Szadzińska, J.; Kowalski, S.J. Microwave- and Infrared-Assisted Convective Drying of Green Pepper: Quality and Energy Considerations. *Chem. Eng. Process. Process Intensif.* **2015**, *98*, 155–164. [CrossRef]
24. Gamboa-Santos, J.; Montilla, A.; Cárcel, J.A.; Villamiel, M.; García-Pérez, J.V. Air-Borne Ultrasound Application in the Convective Drying of Strawberry. *J. Food Eng.* **2014**, *128*, 132–139. [CrossRef]
25. Kowalski, S.J.; Pawłowski, A.; Szadzińska, J.; Łechtańska, J.; Stasiak, M. High Power Airborne Ultrasound Assist in Combined Drying of Raspberries. *Innov. Food Sci. Emerg. Technol.* **2016**, *34*, 225–233. [CrossRef]
26. Musielak, G.; Mierzwa, D.; Kroehnke, J. Food Drying Enhancement by Ultrasound—A Review. *Trends Food Sci. Technol.* **2016**, *56*, 126–141. [CrossRef]
27. de la Fuente-Blanco, S.; Riera-Franco de Sarabia, E.; Acosta-Aparicio, V.M.; Blanco-Blanco, A.; Gallego-Juárez, J.A. Food Drying Process by Power Ultrasound. *Ultrasonics* **2006**, *44*, e523–e527. [CrossRef] [PubMed]
28. Mujumdar, A.S. Principles, Classification, and Selection of Dryers. In *Handbook of Industrial Drying*, 4th ed.; Mujumdar, A.S., Ed.; CRC Press: Boca Raton, FL, USA, 2014; pp. 3–30, ISBN 978-1-4665-9665-8.
29. Cui, Z.-W.; Xu, S.-Y.; Sun, D.-W. Microwave–Vacuum Drying Kinetics of Carrot Slices. *J. Food Eng.* **2004**, *65*, 157–164. [CrossRef]
30. Szadzińska, J.; Mierzwa, D.; Musielak, G. Ultrasound-Assisted Convective Drying of White Mushrooms (*Agaricus bisporus*). *Chem. Eng. Process.—Process Intensif.* **2022**, *172*, 108803. [CrossRef]
31. Pathare, P.B.; Opara, U.L.; Al-Said, F.A.J. Colour Measurement and Analysis in Fresh and Processed Foods: A Review. *Food Bioprocess Technol.* **2013**, *6*, 36–60. [CrossRef]
32. *Origin(Pro)*; OriginLab Corporation: Northampton, MA, USA, 2024.
33. May, B.K.; Perré, P. The Importance of Considering Exchange Surface Area Reduction to Exhibit a Constant Drying Flux Period in Foodstuffs. *J. Food Eng.* **2002**, *54*, 271–282. [CrossRef]
34. Wang, Y.; Zhang, M.; Mujumdar, A.S. Microwave-Assisted Drying of Foods—Equipment, Process and Product Quality. In *Modern Drying Technology, Volume 5: Process Intensification*; Tsotsas, E., Mujumdar, A.S., Eds.; Wiley-VCH Verlag GmbH & Co. KGaA: Weinheim, Germany, 2014; Volume 5, pp. 279–315, ISBN 978-3-527-63172-8.
35. Arslan, A.; Soysal, Y.; Keskin, M. Comparative Investigation of Drying and Quality Characteristics of Organic and Conventional Black Carrots Dried by Intermittent Microwave and Hot Air. *Tekirdağ Ziraat Fakültesi Derg.* **2023**, *20*, 591–604. [CrossRef]
36. Arslan, A.; Soysal, Y.; Keskin, M. Comparing Hot Air Drying Kinetics and Color Quality of Organic and Conventional Sweet Red Peppers. *Mustafa Kemal Üniversitesi Tarım Bilim. Derg.* **2020**, *25*, 271–283. [CrossRef]
37. Kroehnke, J.; Szadzińska, J.; Stasiak, M.; Radziejewska-Kubzdela, E.; Biegańska-Marecik, R.; Musielak, G. Ultrasound- and Microwave-Assisted Convective Drying of Carrots—Process Kinetics and Product's Quality Analysis. *Ultrason. Sonochem.* **2018**, *48*, 249–258. [CrossRef]
38. Kowalski, S.J.; Mierzwa, D.; Stasiak, M. Ultrasound-Assisted Convective Drying of Apples at Different Process Conditions. *Dry. Technol.* **2017**, *35*, 939–947. [CrossRef]
39. Mierzwa, D.; Szadzińska, J. The Microwave-Assisted Convective Drying of Kale (*Brassica oleracea* L. Var. *Sabellica* L.) Using Continuous and Changeable Power Radiation. *J. Food Process Eng.* **2019**, *42*, e13004. [CrossRef]
40. Kroehnke, J.; Musielak, G. Ultrasound-Assisted Drying of Apples—Process Kinetics, Energy Consumption, and Product Quality. *Dry. Technol.* **2024**, *42*, 1757–1765. [CrossRef]
41. Pittia, P.; Antonello, P. Chapter 2—Safety by Control of Water Activity: Drying, Smoking, and Salt or Sugar Addition. In *Regulating Safety of Traditional and Ethnic Foods*; Prakash, V., Martín-Belloso, O., Keener, L., Astley, S., Braun, S., McMahon, H., Lelieveld, H., Eds.; Academic Press: San Diego, CA, USA, 2016; pp. 7–28, ISBN 978-0-12-800605-4.
42. Pham, N.D.; Karim, M.A. Investigation of Nutritional Quality Evolution of Papaya during Intermittent Microwave Convective Drying. *Dry. Technol.* **2022**, *40*, 3694–3707. [CrossRef]
43. Bae, S.-H.; Jeong, M.-G.; Kim, J.-H.; Lee, W.-S. A Continuous Power-Controlled Microwave Belt Drier Improving Heating Uniformity. *IEEE Microw. Wirel. Compon. Lett.* **2017**, *27*, 527–529. [CrossRef]
44. Mierzwa, D.; Szadzińska, J.; Pawłowski, A.; Pashminehazar, R.; Kharaghani, A. Nonstationary Convective Drying of Raspberries, Assisted by Microwaves and Ultrasound. *Dry. Technol.* **2019**, *37*, 988–1001. [CrossRef]

applied sciences

MDPI

Article

Impact of High-Pressure Processing on Quality and Safety of High-Oil-Content Pesto Sauce: A Comparative Study with Thermal Processing

Ehsan Shad [1], Kaisa Raninen [1], Svetlana Podergina [2], Lok In Chan [3], Kam Pui Tong [3], Heidi Hälikkä [1], Marjo Huovinen [2] and Jenni Korhonen [1,*]

[1] Institute of Public Health and Clinical Nutrition, University of Eastern Finland, 70210 Kuopio, Finland; ehsan.shad@uef.fi (E.S.); kaisa.raninen@uef.fi (K.R.); heidi.halikka@uef.fi (H.H.)

[2] School of Pharmacy, Faculty of Health Sciences, University of Eastern Finland, 70210 Kuopio, Finland; marjo.huovinen@uef.fi (M.H.)

[3] School of Biological Sciences, University of Hong Kong, Pokfulam Road, Hong Kong SAR 999077, China; kinkinkichan@gmail.com (L.I.C.); isaa@connect.hku.hk (K.P.T.)

* Correspondence: jenni.korhonen@uef.fi

Abstract: High-pressure processing (HPP) is a promising technology for increasing the shelf life of food, with minimal effects on the nutritional or sensory quality. However, there has been a concern that high-oil-content foods may protect food pathogens in HPP, and that HPP can affect the quality of lipids. We inoculated *Listeria monocytogenes* and *Salmonella* Typhimurium into 34% and 54% oil-content pesto sauce, processed them either with HPP (600 MPa, 4 min) or thermal processing (82 °C, 5 min), and analyzed bacteria counts, pH, GC-MS (Terpene compounds), the time–kill kinetic study, and lipid oxidation value for 60 days in refrigerating storage (5 ± 2 °C). Our findings show that HPP significantly reduced the number of bacteria (more than 4-log) compared to thermal processing or non-processing. Additionally, we discovered terpene compounds (highest-level terpene: L-linalool, eugenol, and 1,8-cineol) in pesto oil that exhibit antimicrobial activity. Different oil content did not have any significant effect on bacteria levels. Regarding chemical results, all samples were of acceptable quality, and the processes did not show any negative effect on lipid oxidation (Peroxide and P-Anisidine value under 10 meq per kilogram of oil). In conclusion, our study indicates that HPP is a suitable method for high-oil-content pesto sauce. In addition, functional compounds naturally present in pesto may contribute to maintaining its microbial and chemical quality.

Keywords: high-pressure processing; pesto sauce; shelf life; microbiological quality; lipid oxidation

1. Introduction

With the media giving much coverage to health, consumers today are more aware of health aspects and interested in the quality of life. Understanding has risen as to the importance of safe and quality food among consumers, to enhance their own well-being, as well as reduce the cases of diseases such as diabetes and cardiac disorders due to improper lifestyles [1]. Due to what consumers want, the quality of food now relates to food security, nutrition properties, sensory characteristics, and sustainability [2]. In this result, the connection between quality and healthiness of food products has gathered power [3]. Considering the wishes of consumers, some non-thermal methods of processing were brought, to obtain high-quality, safe, and nutritional food products [4,5]. However, Conventional thermal processing methods usually compromise the nutritional integrity, taste, and texture of foods by exposure to high temperatures. Non-thermal methods such as HPP, pulsed electric fields, and ultrasound have great potential for preserving heat-labile nutrients, extending the shelf life of food, and demonstrating higher environmental friendliness [6,7]. Non-thermal processing is a promising approach in fulfilling the requirements of advanced food safety and sustainable development, matching consumers' needs at the present time [8,9].

HPP is a processing method that kills potential pathogens through the application of high pressure, rather than heat, to food [10]. The processing is typically carried out in sales packaging. HPP is conventionally applied to liquid and solid foods with high water content, over 87%, like ready-to-eat meals, vegetables, fruits, and dairy products [11,12]. HPP is uniform and instantaneous, unlike thermal processes [13]. The pressure transmission is independent of both mass and time, and, for this reason, the process is short, and minimal nutrient change occurs in food products. Furthermore, the scalability of this technique may be easily and safely transferred from the laboratory to the industry, because HPP is not reliant on the product geometry and size [14].

The European Food Safety Authority (EFSA) has evaluated the safety and efficacy of HPP and believes it is capable of reaching adequate reductions of harmful microorganisms without introducing other safety risks [15]. Since HPP does not denature covalent bonds, it preserves the properties of food products; hence, only the taste, texture, and nutritional value are minimally affected. The technology is active on pathogens such as *Listeria* spp., *Salmonella* spp., and *Brucella* spp., besides proteins associated with microbial replication [16,17]. The improvement in food safety includes the inhibition of essential enzyme synthesis, damage to the microbial cell membrane, disruption of ion exchange and permeability, and leakage of cellular content [17,18]. However, it is important to underline the fact that the efficacy of HPP should be measured in various food matrices.

HPP is currently utilized for a variety of products. In addition to deactivating microorganisms, it is crucial to maintain the stability of other compounds, particularly fats, after undergoing the process. Some studies have reported that a high lipid content in food materials, such as peanut butter and cheese, increases resistance against spoilage and pathogenic bacteria. This is due to the low water activity rendered by lipids [19,20]. More recently, lipid fractions have been the focus of research on how they respond to HPP, since harmful secondary products of oxidation can inflict very negative impacts on the final product [21]. Numerous studies have demonstrated that high-pressure processing in high-fat foods leads to increased lipid oxidation [22,23]. A study on yak fat revealed that applying 600 Mbar pressure significantly increased secondary oxidation and caused changes in the amount of fatty acids, particularly unsaturated fatty acids [24]. Additionally, various studies on lipid oxidation after high-pressure processing in food products containing fat and high fat, such as pork meat and fat [25], salmon [26], cold-smoked salmon [27] and beef muscle [28] have indicated that oxidation can occur at pressures ranging from 200 to 600 Mbar, depending on the nature of the food. The compression resulting from the HPP can increase the temperature of the food by 3 °C per 100 Mbar [29], and, in high-fat food, the increase is approximately 8 to 9 °C [30]; after the decompression, the food's temperature returns to its initial level. This temperature rise, along with the composition of fatty acids and time of process, can influence the oxidation rate of lipids in high-fat products [21].

Pesto sauce is a flavorful sauce that is employed in variety of dishes. Its main ingredients are basil, along with cheese, oil, and nuts. The unique taste and aroma of pesto come from basil. This distinct aroma and flavor are due to terpene compounds in basil, such as methyl cinnamate, linalool, eugenol, eucalyptol, and estragole [31,32]. These compounds not only create the aroma and flavor in basil but also have beneficial properties, such as antimicrobial and antioxidant effects [33]. Several studies reported that the antioxidant and antimicrobial potential of essential oil is related to the presence of terpene and aromatic compounds [34].

The aim of the study will be to investigate the combined impact of high-pressure processing (HPP) and formulation ingredients on the microbial and chemical properties of pesto sauce with high oil content. It is important to understand whether high oil content can protect bacteria and change HPP efficacy or not, as well as how HPP can change lipid quality in pesto sauce with functional ingredients (basil).

2. Materials and Methods

2.1. Sample Preparation

The pesto sauces (Välimeren Herkut Pesto-Kastike, including rape oil, basil (41.5%), cheese, olive oil, sunflower seeds, salt, sugar, garlic, black pepper, ascorbic acid, and potassium sorbate) were processed with 34% and 54% oil content in the Välimeren Herkut factory of Nurmijärvi, Finland, and delivered to the laboratory of the university under cold-chain conditions of 5 ± 2 °C. The samples were inoculated with bacteria (*Listeria monocytogenes* ATCC 4342 and *Salmonella enterica* subsp. enterica serovar Typhimurium (*Salmonella* Typhimurium) ATCC 13311) (ATCC, Manassas, VA 20110, USA) and repacked in 50 mL HPP-tested PET bottles (50 g pesto sauce) for the next step, and some were repacked without bacteria to evaluate processing properties. All samples were divided according to the processing method into three groups: non-processed (NP), thermal processed (TP), and HPP (HP). The samples of the HPP-treated group were transferred to the Toripiha factory (Suonenjoki, Finland) under conditions of the cold chain (5 ± 2 °C) for the process of HPP (600 MPa, 4 min). After the process was completed, the samples were taken back to the laboratory of the university for the examination of microbial and chemical features. At the university laboratory, the water-bath thermal processing was carried out at 82 °C for 5 min. A thermometer within the paralleled samples assured that the expected temperature had been reached. Finally, all the samples were then stored under cold conditions (5 ± 2 °C) for 60 days, to evaluate their storage stability and properties.

2.2. Bacteria Inoculation

The *L. monocytogenes* and *S.* Typhimurium were inoculated into the Tryptic Soy Broth medium (84675.0500, VWR, Avantor, Aurora, OH, USA) and then incubated (B8000, Termaks, Upplands Väsby, Sweden) for 24 h at 37 °C to activate the cells. The inoculum suspension was standardized using the 0.5 McFarland standard method with a spectrophotometer (UV-1600PC, VWR, Aurora, OH, USA) [35]. Then, the bacteria were prepared in amounts that would be added to the pesto sauces, which aimed for approximately 6–7 log CFU/g [36]. The inoculated samples were designated for making pesto into the following groups: LNP (pesto prepared with *L. monocytogenes*, non-processed), LTP (pesto prepared with *L. monocytogenes*, thermal processed), LHP (pesto prepared with *L. monocytogenes*, HPP processed), SNP (pesto prepared with *S.* Typhimurium, non-processed), STP (pesto prepared with *S.* Typhimurium, thermal processed), and SHP (pesto prepared with *S.* Typhimurium). The samples without inoculated bacteria were named NP, TP, or HP according to the processing method.

2.3. Microbiological Analysis

2.3.1. Bacteria Enumeration

To assess the microbiological quality of the pesto sauce, bacterial counts were determined by placing 10 g of pesto sauce in 90 mL of sterile 0.9% NaCl (7647-14-5, Fisher Chemical, Roskilde, Denmark) solution and mixing it for 1 min in a Stomacher blender (Stomacher 400 circulator, Seward, Easting CI, UK) at 260 rpm. After homogenization, additional decimal dilutions were prepared using 0.9% NaCl solution as the diluent. Subsequently, 100 μL of the final three dilutions was transferred to ready-to-use Petri-dish media (82.1194.500, SARSTEDT, Hanover, Germany). All counts were expressed as log CFU per gram and were obtained in duplicate. *L. monocytogenes* and *S.* Typhimurium counts were determined using Listeria Selective Agar base Ottaviani and Agosti (LSA) (84748.0500, VWR, Avantor, Aurora, OH, USA) and Xylose Lysine Deoxycholate agar (XLD) (84667.0500, VWR, Avantor, Aurora, OH, USA), respectively, with incubation at 37 °C for 48 h.

2.3.2. Time–Kill Kinetic Studies

The time–kill kinetic study [37] was carried out with 500 μL oil separated from pesto sauce (the oil was separated from the solid materials of the pesto by simply centrifuging at 4000 rpm for 5 min (5415D, Eppendorf, Hamburg, Germany)) in 10 mL Tryptic soy broth medium (84675.0500, VWR, Avantor, USA), which was inoculated (0.5 McFarland standard method, using spectrophotometer (UV-1600PC, VWR, USA)) [35] with *L. monocytogenes* and *S.* Typhimurium, separately. Samples were taken at different time points (0, 2, 4, 6, and 24 h) during the experiment. Each sample was 1 mL in volume and was mixed with a saline solution to create serial dilutions. After dilution, 100 µL from each dilution was placed onto specific mediums (LSA and XLD medium) and then kept in an incubator for 24 h at 37 °C. This process was repeated three times, to ensure accuracy. The growth rate was then calculated before and after treatment with oil, and expressed as log CFU/mL.

2.4. Chemical Analysis

2.4.1. GC/MS Analysis

Terpenes in the oil phases of pestos were monitored with gas chromatography–mass spectrometry (GC-MS). Each sample was analyzed as three replicants. The oil was separated from the solid materials of the pesto by simply centrifuging at 4000 rpm for 5 min (5415D, Eppendorf, Germany) and pipetting. Then pesto oil samples were diluted with hexane at a 1:10 ratio, the internal standard (99% 1-chlorooctane, Alfa Aesar/Thermo, Thermo Scientific Chemicals, Haverhill, Boston, MA, USA) was added to the samples at a $1{:}10^5$ ratio, and the samples were filtered using Whatman Mini-UniPrep G2 Syringeless Filters (GE HealthCare, Amersham, UK). The analysis was performed with a GC-MS device consisting of GCMS-QP2010 Ultra and AOC-5000 Plus injection system (Shimadzu Scientific Instruments, Columbia, MD, USA). A 1 µL sample was injected into the injector at 280 °C temperature and split 5. Separation of compounds was performed with BP-5 column (30 m/0.25 mm/0.25 µm, Trajan Scientific and Medical, RingWood, Australia) using oven temperature 40–280 °C and helium (N60, Woikoski, Mäntyharju, Finland) as a carrier gas with linear velocity 39.5. The temperatures of ion source and interface were 200 and 290 °C, respectively, and ions were scanned at 33–400 m/z. The compounds were tentatively identified by comparing their mass spectra with data from mass spectral libraries (NIST 11 Mass Spectral library, The National Institute of Standards and Technology, USA Wiley Registry 10th Edition, John Wiley & Sons, USA and Flavour & Fragrance Natural & Synthetic Compounds GCMS library FFNSC 2, Shimadzu Corp, Kyoto, Japon) and checking their retention indexes with alkane standards (C8-C40 Alkanes Calibration Std, Supelco, Bellefonte, PA, USA). The areas of the GC-MS signal intensities of the terpenes were integrated in GCMSsolution software version 4.20 (Shimadzu, Scientific Instruments, Haverhill, Boston, USA) to check if the processes affected the levels of terpenes in pesto oils.

The coefficient of variation of the GC-MS method was monitored with the internal standard, and it was 16.0% as calculated from all the samples in the GC-MS ($n = 18$). Two replicant samples were excluded from the analysis, since their signal levels varied more than 30% from those of the other two replicants and were therefore considered as outliers.

2.4.2. pH

A 2 g sample of pesto sauce was mixed with 8 mL of deionized water for 2 min. The pH was then measured at room temperature using a digital pH meter (CG 842, SCHOTT, Mainz, Germany) with a probe-type combined electrode (Ingold, Amsterdam, The Netherlands) by immersing the electrode directly into the mixture.

2.4.3. Total Phenolic Content (TPC)

Two mL vials containing 0.5 g of frozen pesto sauce samples were spiked with internal standards and homogenized with methanol (20864.320, VWR, Avantor, USA) (1.3 mL) for 2 min in a Tissulyzer LT homogenizer (85600, Qiagen, Venlo, The Netherlands). The obtained extracts were centrifuged at 15,000× *g* for 10 min at 4 °C using a 5424R centrifuge (Eppendorf, Hamburg, Germany) [38]. The supernatant collected was used to determine the TPC, based on the method by Hilma et al. (2018), but with some modifications [39]. Briefly, 100 µL of supernatant was mixed with 100 µL of 0.1 N Folin–Ciocalteu reagent (1090010100, Merck, Darmstadt, Germany) in a 2 mL vial and incubated for 1 min at room temperature in the dark. A total of 500 μL of 7% Na_2CO_3 solution (31432, Riedel-de Haën, Hanover, Germany) and 1.3 mL of deionized water were added to the mixture. The mixture was additionally incubated at room temperature in darkness for 120 min before taking absorbance at 750 nm. Phenolic content is obtained from the absorption against the standard curve previously developed (Supplementary S1. TPC content was expressed in milligrams of Gallic Acid (398225-100, Sigma-Aldrich, Steinheim, Germany) Equivalents (GAEs) per 100 g of pesto sauce.

2.4.4. Total Antioxidant Capacity

Antioxidant activity percentage (AA%) was evaluated according to Kulkarani and Aradhya (2005) methods, with some modification [40]. A 0.1 mL sample (same sample extract in TPC) was mixed with 0.9 mL methanol and, after that, mixed with 1 mL 2,2-diphenyl-1-picrylhydrazil (DPPH) (0.500 μM in methanol). The control sample was prepared in a similar way, and the sample was replaced with water. The solution was mixed very well and kept in dark conditions for 30 min. The absorbance of the final solution was determined at 517 nm by a spectrophotometer. The antioxidant activity was calculated using the following equation:

$$AA\% = \left[1 - \frac{Asample\ (517\ nm)}{Acontrol\ (517\ nm)}\right] \times 100$$

2.4.5. Lipid Oxidation

The present study assessed the peroxide value as an indicator of initial oxidation and the p-Anisidine value as an indicator of secondary oxidation [41]. To determine the peroxide value of pesto sauce after various processing methods, the oil part was first separated by centrifuging (5415D, Eppendorf, Germany) control samples at 4000 rpm for 5 min and keeping the oil at 4–6 °C. Chemical solutions for the titration method were prepared, including an acetic acid–chloroform solution, saturated potassium iodide (7681-11-0, Merck, Germany) solution, 0.1 M sodium thiosulfate (7772-98-7, Sigma-Aldrich, Germany) solution, 10% sodium lauryl sulfate (Riedel-de Haën, Hanover, Germany), and 1% starch (9005-84-9, Thermo Fisher, Rochester, New York, NY, USA) indicator solution. The titration technique required weighing 5 g of extracted oil, dissolving it in a mixture of acetic acid (64-19-7, VWR, Amsterdam, The Netherlands) and chloroform (64-19-7, VWR, Amsterdam, The Netherland), and adding potassium iodide solution. The mixture was then titrated using sodium thiosulfate until a color change appeared. To liberate all iodine from the chloroform layer, sodium lauryl sulfate and starch indicator were added, and then titrated until the blue color disappeared [42]. The volume of titrant used in both blank and control samples was recorded, and the peroxide value was determined using the following equation:

$$\text{Peroxide value (milliequivalents / peroxide /1000 } g \text{ test portion)} = \frac{(S - B) \times M \times 1000}{Mass\ of\ test\ portion\ (g)}$$

S = Volume of titrant used for control samples (mL)

B = Volume of titrant used for blank (mL)

M = Molarity of sodium thiosulfate ($Na_2S_2O_3$) standard solution

To determine the p-anisidine content of the pesto sauce control sample, p-anisidine reagent (S8426458346, Sigma-Aldrich, Germany) was made by dissolving 0.25 g of p-anisidine in 100 mL of glacial acetic acid (64-19-7, VWR, The Netherlands), protected from light with aluminum foil. For the measurement, 4 g of the oil was diluted with isooctane in a 25 mL volumetric flask, and then 1 mL of this solution was diluted with 9 mL isooctane (540-84-1, Merck, Germany). The absorbance (A_b) of this solution was measured at 350 nm using UV–Visible spectroscopy, with isooctane serving as the blank. The oil solution and isooctane were then combined with 1 mL of p-anisidine reagent in separate test tubes, agitated, and left for 10 min. The absorbance (A_s) of the oil solution was measured again at 350 nm, using the isooctane mixture as the blank. The p-anisidine value was eventually obtained using the following equation [43]:

$$\text{p-anisidine value} = 25 \times (1.2A_s - A_b) \times \frac{dilution\ factor}{mass\ of\ oil\ portion\ (g)}$$

2.5. Statistical Analysis

All experiments were carried out in triplicate. One-way analysis of variance (ANOVA) and the Duncan group were used to determine the significance of differences ($p < 0.05$) between means. GraphPad Prism 10 was used for the statistical analysis.

3. Results

3.1. Bacteria Enumeration Results of the Pesto Sauces

The results of the *L. monocytogenes* and *S.* Typhimurium counts are shown in Figure 1. The result of *L. monocytogenes* counts before and after processing is reported in Figure 1a. HPP drastically ($p < 0.05$) decreased the *L. monocytogenes* count in the LHP-34 (2.37 ± 0.18 log CFU/g) and LHP-54 (2.35 ± 0.17 log CFU/g) samples compared to the non-processed samples LNP-34 (6.78 ± 0.15 log CFU/g) and LNP-54 (6.91 ± 0.10 log CFU/g). There was no significant difference between HPP samples with different oil content in the amount of *L. monocytogenes*. Also, the thermal processing led to a significant decrease ($p < 0.05$) in *L. monocytogenes* counts in LTP-34 (2.90 ± 0.46 log CFU/g) and LTP-54 (4.99 ± 0.04 log CFU/g) compared to LNP samples. The oil content of pesto sauce was found to impact the survival rate of *L. monocytogenes* in these samples. In other words, the LTP-34 samples had a significant ($p < 0.05$) reduction in *L. monocytogenes* compared to the LTP-54 samples (Figure 1a). Figure 1b shows *L. monocytogenes* counts during the storage time (60 days). The results showed that *L. monocytogenes* drastically decreased more than 5-fold log after 30 days in non-processed samples. In contrast, LHP and LTP-34 samples revealed no significant difference during storage time (Figure 1b).

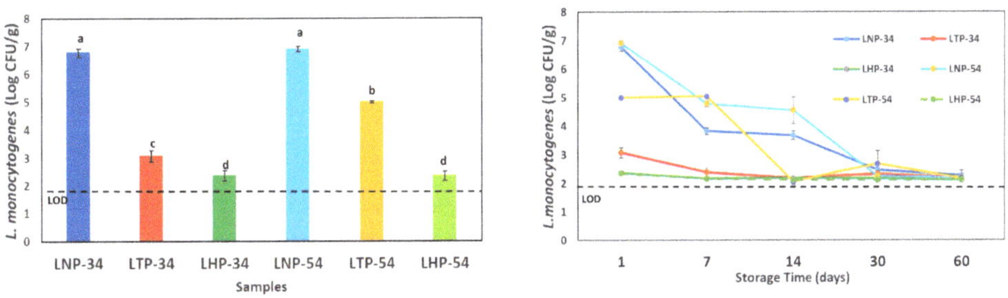

(a) (b)

Figure 1. *Listeria monocytogenes* counts in the pesto samples having 34 or 54% oil content and different processing methods on day 1 (**a**) and during 60 days of storage (**b**). Data represent the mean ± standard deviation of three independent replicates; different superscript letters in each column indicate significant differences ($p < 0.05$). LNP: non-processed samples with *L. monocytogenes*, LTP: thermal-processed samples with *L. monocytogenes*, LHP: HPP samples with *L. monocytogenes*. LOD: limit of detection (2 log CFU/g).

S. Typhimurium counts after exposure to HPP and thermal processing were below the limit of detection (LOD, 2 log CFU/g), being significantly different compared to non-processed samples (SNP-34: 5.07 ± 0.35, SNP-54: 5.06 ± 0.18 Log CFU/g) on day 1. Also, the results showed that *S.* Typhimurium counts were under LOD for 60 days in processed samples. Surprisingly, the *S.* Typhimurium counts decreased under the LOD after 7 days, also in non-processed samples. The oil content did not affect the survival rate of *S.* Typhimurium.

3.2. Time–Kill Kinetic Study Results of the Pesto Sauces

In Table 1, the results of the time–kill study of *L. monocytogenes* are presented. The results indicated that there was no statistically significant ($p < 0.05$) difference between the samples control, NPO-34, and NPO-54 at 0 h, and this finding persisted at 2 and 4 h. However, at 4 h, the growth rate of *L. monocytogenes* in NPO-34 (6.66 ± 0.23 log CFU/g) was significantly lower than in the control (7.62 ± 0.12 log CFU/g) and NPO-54 (7.49 ± 0.19 log CFU/g). As the incubation time increased to 24 h, no significant difference was observed among the samples. Also, the same trend ($p < 0.05$) was observed in TPO and HPO groups, and the samples containing oil from pesto sauce with 34% oil showed less growth at the fourth point of the day compared to their peers, but, with the continuation of the incubation process (24 h), no significant ($p < 0.05$) difference was observed.

Table 1. Time–kill kinetic study results of oils from different pesto sauces against *L. monocytogenes*.

Time (Hour)	0	2	4	6	24
Treatment			***L. monocytogenes* Counts (log CFU/mL)**		
Control	4.82 ± 0.12 ^{Ad}	6.14 ± 0.29 ^{Ac}	7.62 ± 0.12 ^{Ab}	8.18 ± 0.34 ^{Aa}	8.28 ± 0.25 ^{Aa}
NPO-34	4.93 ± 0.17 ^{Ad}	6.05 ± 0.31 ^{Ac}	6.66 ± 0.23 ^{Bb}	7.77 ± 0.27 ^{Aa}	8.02 ± 0.14 ^{Aa}
NPO-54	4.94 ± 0.35 ^{Ad}	6.07 ± 0.16 ^{Ac}	7.49 ± 0.19 ^{Ab}	8.07 ± 0.19 ^{Aa}	8.20 ± 0.33 ^{Aa}
Control	4.89 ± 0.27 ^{Ae}	6.25 ± 0.11 ^{Ad}	7.51 ± 0.36 ^{Ac}	8.10 ± 0.18 ^{Ab}	8.73 ± 0.24 ^{Aa}
TPO-34	4.85 ± 0.08 ^{Ae}	6.10 ± 0.29 ^{Ad}	6.73 ± 0.24 ^{Bc}	7.82 ± 0.36 ^{Ab}	8.50 ± 0.26 ^{Aa}
TPO-54	4.90 ± 0.19 ^{Ad}	6.13 ± 0.38 ^{Ac}	7.39 ± 0.15 ^{Ab}	8.12 ± 0.31 ^{Aa}	8.64 ± 0.21 ^{Aa}
Control	4.85 ± 0.26 ^{Ad}	6.11 ± 0.42 ^{Ac}	7.63 ± 0.17 ^{Ab}	8.21 ± 0.26 ^{Aa}	8.73 ± 0.21 ^{Aa}
HPO-34	4.92 ± 0.17 ^{Ae}	6.05 ± 0.14 ^{Ad}	6.65 ± 0.31 ^{Bc}	7.75 ± 0.45 ^{Ab}	8.50 ± 0.09 ^{Aa}
HPO-54	4.93 ± 0.21 ^{Ad}	6.08 ± 0.28 ^{Ac}	7.52 ± 0.20 ^{Ab}	8.09 ± 0.19 ^{Aa}	8.64 ± 0.38 ^{Aa}

At the same time of storage, different capital letters indicate significant ($p < 0.05$) differences between different samples. For the same sample, different lowercase letters indicate significant ($p < 0.05$) differences over time. Control: bacteria without oil, NPO-34: bacteria with oil from 34% oil content non-process sample, NPO-54: bacteria with oil from 54% oil-content non-processed sample, TPO-34: bacteria with oil from 34% oil-content thermal-processed sample, TPO-54: bacteria with oil from 54% oil-content thermal-processed sample, HPO-34: bacteria with oil from 34% oil-content HPP sample, HPO-54: bacteria with oil from 54% oil-content HPP sample.

The results of *S.* Typhimurium growth are presented in Table 2. The data indicate that, initially, there is no significant difference in the growth of *S.* Typhimurium in the control, TPO, and HPO samples. However, after 2 h of incubation, there is a notable ($p < 0.05$) decrease in the growth rate of *S.* Typhimurium in the samples containing oil from pesto sauce with 34% oil, compared to the other samples. Subsequently, there were no significant differences observed in the growth rates among the different samples over the continued duration of incubation.

Table 2. Time–kill kinetic study results of oils from different pesto sauces against *S.* Typhimurium.

Time (Hour)	0	2	4	6	24
Treatment			***S.* Typhimurium Counts (log CFU/g)**		
Control	6.16 ± 0.21 ^{Ac}	6.50 ± 0.08 ^{Ab}	8.07 ± 0.34 ^{Aa}	8.25 ± 0.26 ^{Aa}	8.55 ± 0.22 ^{Aa}
NPO-34	6.15 ± 0.32 ^{Ab}	5.90 ± 0.22 ^{Bb}	7.89 ± 0.25 ^{Aa}	8.22 ± 0.42 ^{Aa}	8.74 ± 0.16 ^{Aa}
NPO-54	6.04 ± 0.12 ^{Ac}	6.75 ± 0.13 ^{Ab}	7.96 ± 0.41 ^{Aa}	8.10 ± 0.25 ^{Aa}	8.63 ± 0.21 ^{Aa}
Control	5.98 ± 0.15 ^{Ac}	6.67 ± 0.14 ^{Ab}	7.86 ± 0.29 ^{Ab}	8.37 ± 0.19 ^{Aa}	8.69 ± 0.33 ^{Aa}
TPO-34	6.12 ± 0.22 ^{Ab}	6.02 ±0.26 ^{Bb}	8.02 ± 0.09 ^{Aa}	8.21 ± 0.32 ^{Aa}	8.58 ± 0.24 ^{Aa}
TPO-54	6.06 ± 0.14 ^{Ac}	6.58 ± 0.14 ^{Ab}	7.94 ± 0.17 ^{Aa}	8.29 ± 0.34 ^{Aa}	8.47 ± 0.31 ^{Aa}

Table 2. Cont.

Time (Hour)	0	2	4	6	24
Treatment	S. Typhimurium Counts (log CFU/g)				
Control	6.16 ± 0.11 Ac	6.51 ± 0.17 Ab	7.95 ± 0.31 Aa	8.30 ± 0.14 Aa	8.46 ± 0.17 Aa
HPO-34	6.33 ±0.15 Ab	5.90 ± 0.10 Bc	7.87 ±0.17 Aa	8.22 ± 0.31 Aa	8.67 ± 0.26 Aa
HPO-54	6.36 ±0.19 Ab	6.67 ± 0.16 Ab	7.93 ±0.24 Aa	8.25 ± 0.28 Ab	8.73 ± 0.18 Aa

At the same time of storage, different capital letters indicate significant ($p < 0.05$) differences between different samples. For the same sample, different lowercase letters indicate significant ($p < 0.05$) differences over time. Control: bacteria without oil, NPO-34: bacteria with oil from 34% oil-content non-processed sample, NPO-54: bacteria with oil from 54% oil-content non-processed sample, TPO-34: bacteria with oil from 34% oil-content thermal-processed sample, TPO-54: bacteria with oil from 54% oil-content thermal-processed sample, HPO-34: bacteria with oil from 34% oil-content HPP sample, HPO-54: bacteria with oil from 54% oil-content HPP sample.

3.3. GC-MS Results of the Pesto Oils

A total of 12 terpene compounds were identified in the pesto oil samples with GC-MS (Table 3). Specifically, L-linalool, eugenol, and 1,8-cineol had the highest levels of terpenes in the samples. Interestingly, it was found that the HPP or thermal processes did not affect the levels of terpenes in the pesto oils. Additionally, the terpene levels were lower in the 54% oil samples compared to the 34% oil samples. This could be due to the higher oil content in the 54% pesto, which diluted the terpenes more, compared to the 34% oil content.

Table 3. Identified terpenes and their levels in studied basil pesto oils.

				Signal Intensity					
				34% Oil			54% Oil		
Compound Name	SI	RT (min)	RI	NP	TP	HPP	NP	TP	HPP
α-pinene	96	9.28	932	1.14×10^5	1.00×10^5	0.97×10^5	5.55×10^4	5.73×10^4	6.57×10^4
sabinene	90	9.50	974	1.40×10^5	1.48×10^5	1.17×10^5	7.72×10^4	7.54×10^4	8.58×10^4
β-pinene	95	9.67	979	1.78×10^5	1.69×10^5	1.54×10^5	1.02×10^5	0.89×10^5	0.97×10^5
β-myrcene	93	9.96	989	2.22×10^5	2.47×10^5	1.97×10^5	1.26×10^5	1.18×10^5	1.33×10^5
cymene	93	11.16	1029	8.79×10^4	7.91×10^4	8.38×10^4	4.55×10^4	2.42×10^4	3.09×10^4
limonene	92	11.25	1032	1.43×10^5	1.37×10^5	1.28×10^5	6.12×10^4	5.83×10^4	7.45×10^4
1,8-cineol	96	11.36	1035	1.82×10^6	1.68×10^6	1.52×10^6	0.90×10^6	0.82×10^6	0.99×10^6
β-ocimene	93	11.74	1048	1.57×10^5	1.37×10^5	1.28×10^5	7.24×10^4	7.19×10^4	8.66×10^4
l-linalool	96	13.48	1105	9.06×10^6	8.26×10^6	7.51×10^6	4.52×10^6	4.04×10^6	4.96×10^6
l-camphor	92	15.06	1158	0.93×10^5	1.08×10^5	0.71×10^5	4.53×10^4	4.79×10^4	3.62×10^4
α-terpineol	92	16.49	1208	1.75×10^5	1.62×10^5	1.54×10^5	7.96×10^4	6.97×10^4	8.63×10^4
eugenol	93	19.14	1368	5.83×10^6	5.12×10^6	5.12×10^6	2.82×10^6	2.44×10^6	3.07×10^6

SI = similarity index, RT = retention time, RI = retention index, NP = non-processed, TP = thermal processed, HPP = high-pressure processed.

3.4. pH Changes in the Pesto Sauces

Figure 2 illustrates the changes in pH before and after processing pesto sauces. The results indicate a significant increase in pH levels when HPP and thermal processes are used, compared to the non-processed samples.

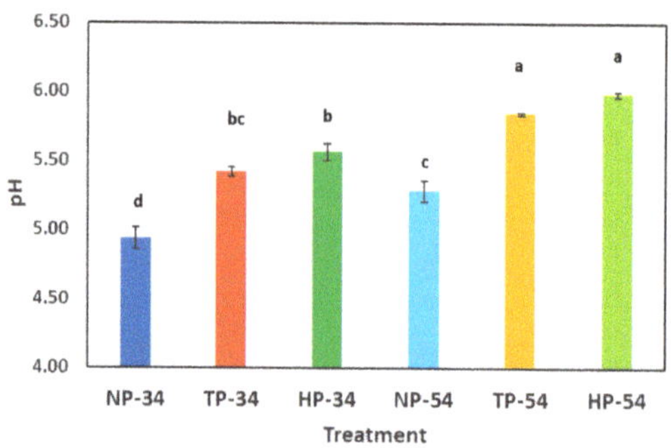

Figure 2. Effect of HPP and thermal processing on pH in pesto sauces. Data represent the mean ± standard deviation of three independent replicates; different superscript letters in each column indicate significant differences ($p < 0.05$). NP-34: 34% oil-content non-processed sample, HPP-34: 34% oil-content HPP sample, TP-34: 34% oil-content thermal-processed sample, NP-54: 54% oil-content non-processed sample, HPP-54: 54% oil-content HPP sample, TP-54: 54% oil-content thermal-processed sample.

3.5. Total Phenolic Content Results of the Pesto Sauces

The TPC results presented in Table 4 show an increase in TPC concentration in the HPP samples compared to the non-processed samples. Additionally, it was observed that the level of TPC in the HPP sample was significantly higher than in the thermal-processed sample, while no significant difference was found between the control and thermal samples. This trend was consistent for both oil content.

Table 4. Effect of HPP and thermal processing on total phenolic content (TPC) and Antioxidant capacity (AA%) in pesto sauces.

Treatment	TPC (mg GAE/100 gr Pesto)	AA%
NP-34	33.01 ± 0.47 ^{e}	61.45 ± 0.79 ^{a}
TP-34	32.81 ± 0.17 ^{e}	60.07 ± 1.06 ^{a}
HP-34	33.99 ± 0.09 ^{d}	62.93 ± 1.29 ^{a}
NP-54	36.04 ± 0.10 ^{c}	60.35 ± 1.21 ^{a}
TP-54	36.87 ± 0.04 ^{b}	60.89 ± 1.10 ^{a}
HP-54	40.01 ± 0.09 ^{a}	61.09 ± 0.35 ^{a}

Data represents the mean ± standard deviation of three independent replicates; different superscript letters in each column indicate significant differences ($p < 0.05$). NP-34: 34% oil content non-processed sample, HPP-34: 34% oil content HPP sample, TP-34: 34% oil content thermal-processed sample, NP-54: 54% oil content non-processed sample, HPP-54: 54% oil content HPP sample, TP-54: 54% oil content thermal-processed sample.

3.6. Total Antioxidant Capacity of Pesto Sauces

The result of antioxidant capacity is shown in Table 4. The result illustrated that there is no significant difference ($p < 0.05$) among treatments, and HPP and thermal processing did not influence the antioxidant activities potential of pesto samples.

3.7. Lipid Oxidation of Pesto Sauces

The peroxide and anisidine values are shown in Figure 3. The peroxide value results indicated that the amount of peroxide in all samples remained lower than the standard (10 milliequivalents of active oxygen per kilogram of oil) during the storage period. However, the peroxide levels in the TP samples increased significantly, compared to the HP and NP samples in 34% and 54% oil content. The analysis of anisidine values showed no noticeable differences among the samples during storage time (60 days) in the cold room.

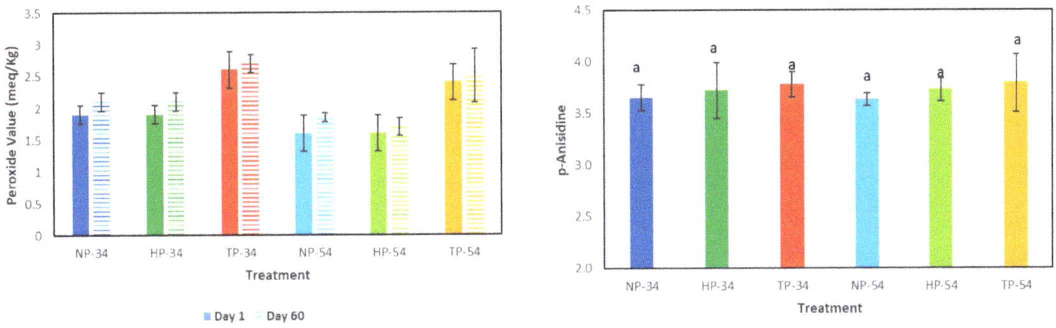

(**a**) (**b**)

Figure 3. Effect of HPP and thermal processing on lipid oxidation in pesto sauce. (**a**) Peroxide value results in day 1 (solid fill) and 60 (pattern fill) of storage time, (**b**) p-anisidine value results after 60 days storage time. Data represent the mean ± standard deviation of three independent replicates; different superscript letters in each column indicate significant differences ($p < 0.05$, Same letters mean no significant differences). NP-34: 34% oil-content non-processed sample, HPP-34: 34% oil-content HPP sample, TP-34: 34% oil-content thermal-processed sample, NP-54: 54% oil-content non-processed sample, HPP-54: 54% oil-content HPP sample, TP-54: 54% oil-content thermal-processed sample.

4. Discussion

HPP is a potential technique for producing high-quality food items with an extended shelf life. However, the applicability of the technique should be tested with various food products. There have been issues with high-oil products that can form a protective barrier against spoilage and pathogenic bacteria during thermal [44,45] and non-thermal processes [46], reducing the efficiency of these procedures in eradicating microorganisms. Furthermore, the high oil concentration might speed up the oxidation of lipids and reduce product quality. This study sought to evaluate the microbiological and chemical aspects of pesto sauce, a high-oil food.

The microbiological study demonstrated that the HPP efficiently reduced the levels of *L. monocytogenes* and *S.* Typhimurium, with no significant influence of oil content on bacterial survival. However, the findings indicated that the presence of oil may alter the efficacy of thermal processing, potentially allowing *L. monocytogenes* bacteria to survive. According to studies, high oil content can reduce the efficacy of thermal processing by lowering water activity (aw). This decrease in aw can lead to *L. monocytogenes* bacteria more resistant to thermal and non-thermal processes [45]. A study by D'Souza et al. (2014) reported that peanut butter formulations could reduce the impact of HPP on *Salmonella* bacteria [47]. The researchers adjusted the aw by adding oil and water, and the results showed that increasing the amount of oil (and therefore lowering aw) dramatically reduced the impact of HPP on Salmonella bacteria. Morales et al. (2006) found that the aw of cheese significantly affected the pressure resistance of *L. monocytogenes* and proposed that fat improved microorganism resistance to HPP destruction in a similar way as thermal inactivation [19]. Cream presents a challenge in removing bacteria, due to its high fat content and reduced aw. Previous study has demonstrated that the microbial load of cream with 35% fat can be significantly reduced with HPP. For example, *Listeria innocua* can be inactivated using HPP at 450 MPa and 25 °C for 10–30 min. It has been suggested that HPP could help extend the refrigerated shelf life of dairy creams [48]. These findings confirm the HPP process as a safe way of producing high-oil-content foods. Evert-Arriagada et al. (2018) found that *Listeria* inactivation increased with higher applied pressure, showing statistical differences based on the strains used (*L. innocua*, *L. monocytogenes CECT 4031* and *L. monocytogenes Scott A*), inoculum levels (4–6 Log CFU/g), and sublethal injury. All three examined strains had the maximum lethality values at 600 MPa; however, *L.*

innocua and *L. monocytogenes* CECT 4031 also had high lethality at 500 MPa. Following treatment, the counts of *L. innocua* and *L. monocytogenes CECT 4031* in fresh cheese gradually increased during cold storage [49]. Another study determined that 6000 bar is the highest effective pressure for inactivating *L. monocytogenes* [10]. Furthermore, in our investigation, all samples containing bacteria showed a significant decrease in *Listeria* and *Salmonella* bacteria during storage. This decrease may be attributable to the antimicrobial compounds extracted from basil during pesto sauce production, and our GC/MS results show that there are 12 different terpene compounds in oils, with linalool being the most abundant. Notably, the oil included terpene compounds with antibacterial characteristics that are naturally found in basil. Basil, a member of the Lamiaceae plant family, is utilized not only as a spice, but is also known for its medical capabilities as a powerful antibacterial, antimutagenic, and chemopreventive agent. According to a study, basil essential oil comprises 65 compounds, with linalool (31.6%) and methyl chavicol (23.8%) being the most prominent [33]. Basil also includes phenolic chemicals, including rosmarinic, chicoric, ferulic, and caffeic acid, which are predominantly phenolic acids. Several studies have demonstrated that basil extract has a high concentration of natural antioxidants with a variety of health advantages, including antibacterial, anti-inflammatory, and antidiabetic activities, due to the synergistic antioxidant impact of these components [50,51].

The findings of the time–kill investigation indicated that these compounds had antimicrobial properties, resulting in a drop in bacteria counts in unprocessed samples throughout storage time. Furthermore, it was shown that the oil content impacted the rate of bacteria reduction, with 34% oil-content samples showing a greater reduction compared to 54% oil-content samples, possibly due to the higher concentration of terpene compounds detected in the GC-MS analysis. Therefore, the higher oil content might have diluted the antimicrobial capabilities of the compounds.

The pH results showed that the pH level mildly increased when thermal and high-pressure processes were applied. This rise can be attributed, for example, to the release of amino from the cheese in the formulation. When thermal and non-thermal processes are used, the protein molecules are broken down and amino compounds are released, which increases the pH level of the products. Ramirez-Suarez et al. (2006) demonstrated that HPP increased the pH in the tuna samples compared to the control. It can occur that pressure, which causes structural changes associated with protein denaturation, can unfold proteins, expose more basic amino acids to the medium, and raise the pH. Several investigations have found that after HPP treatment, the pH of food product samples might rise considerably compared to controls [52,53]. The lipid oxidation analysis in pesto sauce samples revealed that the peroxide and p-anisidine levels were within the standard range. According to the Centre for Food Safety in Hong Kong, the recommended maximum peroxide value in edible fats and oils is 10 meq per kilogram of oil, and the p-anisidine value should be less than 10 meq/kg. After 60 days of cold storage, the oil was found to be of excellent grade. Furthermore, contrary to prior investigations, high pressure and thermal processes did not significantly accelerate lipid oxidation.

The presence of antioxidant-rich phenolic components in pesto sauce inhibits oil oxidation. TPC analysis showed that using basil in the formulation increased the total TPC of all samples [38,54]. Furthermore, the HPP method, in comparison to the thermal process, resulted in increased TPC levels in the samples. This increase in TPC levels may be connected to the improved extractability of some antioxidant components, following HPP [55].

5. Conclusions

Our study demonstrated that HPP improves the microbial and chemical quality of high-oil-content pesto sauce compared with thermal processing. The HPP significantly reduces *Listeria monocytogenes* and *Salmonella* Typhimurium counts (more than 4-log reduction) without being significantly affected by the oil content. Furthermore, the lipids are well preserved by HPP, as shown by low peroxide and p-anisidine values (under 10 meq/Kg oil). This activity increases the total phenolic content, indicating higher antioxidant properties than those observed before the treatment. The presence of functional compounds, particularly phenolics, including terpenes with antimicrobial properties found in basil, likely contributes to the prolonged shelf life and maintains the quality of the pesto sauce. These ingredients—in addition to enhancing the nutritional profile, taste, and texture of the food—help boost the efficacy of this non-thermal preservation tool for high-oil-content foods, further solidifying HPP as an effective method for controlling spoilage organisms.

Supplementary Materials: The following supporting information can be downloaded at: https://www.mdpi.com/article/10.3390/app14209425/s1, Figure S1: Gallic Acid Standard Curve; Table S1. Preparation of gallic acid standards for calibration curve.

Author Contributions: Conceptualization, J.K., E.S. and K.R.; methodology, E.S., K.R., J.K. and S.P.; validation, E.S., K.R. and J.K.; formal analysis, E.S.; investigation, E.S., K.R., S.P., L.I.C., K.P.T. and H.H.: writing—original draft preparation, E.S., K.R., S.P., L.I.C. and K.P.T.: writing—review and editing, J.K. and M.H.; supervision, J.K. and M.H. All authors have read and agreed to the published version of the manuscript.

Funding: This research was partly funded by the European Union, with the European Regional Development Fund A80382, admitted by the Regional Council of Pohjois-Savo.

Institutional Review Board Statement: Not applicable.

Informed Consent Statement: Not applicable.

Data Availability Statement: The raw data supporting the conclusions of this article will be made available by the authors on request.

Acknowledgments: We want to acknowledge the European Regional Development Fund and the Regional Council of Pohjois-Savo, via which the regional ELTUVA project (A80382, 2023–2025) has enhanced research and innovation capacities and the uptake of advanced technologies for developing high-quality and safe food. We want to thank Company Ca-Sa Välimeren Herkut Oy, especially Susanna Työppönen and Petri Klemetti, for preparing and supplying the pesto sauces and being supportive of this study, as well as Toripiha Oy for HPP processing. We thank SIB Labs at Department of Technical Physics in the University of Eastern Finland for providing GC-MS facilities. We are also thankful for Jaana Kapustamäki and Jesse Ojala from Development Company SavoGrow Ltd. for kind cooperation in the ELTUVA project supporting this study. Moreover, we wish to thank our skillful laboratory personnel, Kristiina Kinnunen and Ida Tikkanen, for their assistance in maintaining and developing the laboratory methodology.

Conflicts of Interest: The authors declare no conflicts of interest. The funders had no role in the design of the study; in the collection, analyses, or interpretation of data; in the writing of the manuscript; or in the decision to publish the results. The companies CA-SA Välimeren Herkut and Tori-piha Oy were participating in the ELTUVA project but had no role in the design of the study, analyses, interpretation of data, or writing of the manuscript.

References

1. Rana, J.; Paul, J. Consumer Behavior and Purchase Intention for Organic Food: A Review and Research Agenda. *J. Retail. Consum.* **2017**, *38*, 157–165. [CrossRef]
2. Hosni, H.; Periklis, D.; Baourakis, G. Consumers Attitude towards Healthy Food:"Organic and Functional Foods". *Int. J. Food Beverage Manuf. Bus. Models* **2017**, *2*, 85–99. [CrossRef]
3. Savelli, E.; Murmura, F.; Bravi, L. Healthy and Quality Food Attitudes and Lifestyle: A Generational Cohort Comparison. *TQM J.* **2023**. [CrossRef]
4. Rostamabadi, H.; Nowacka, M.; Colussi, R.; Frasson, S.F.; Demirkesen, I.; Mert, B.; Singha, P.; Singh, S.K.; Falsafi, S.R. Impact of Emerging Non-Thermal Processing Treatments on Major Food Macromolecules: Starch, Protein, and Lipid. *Trends Food Sci. Technol.* **2023**, *141*, 104208. [CrossRef]
5. Zhang, Z.; Wang, L.; Zeng, X.; Han, Z.; Brennan, C.S. Non-thermal Technologies and Its Current and Future Application in the Food Industry: A Review. *Int. J. Food Sci. Technol.* **2019**, *54*, 1–13. [CrossRef]
6. Zhao, Y.-M.; de Alba, M.; Sun, D.-W.; Tiwari, B. Principles and Recent Applications of Novel Non-Thermal Processing Technologies for the Fish Industry—A Review. *Crit. Rev. Food Sci. Nutr.* **2019**, *59*, 728–742. [CrossRef]

7. Jadhav, H.B.; Annapure, U.S.; Deshmukh, R.R. Non-Thermal Technologies for Food Processing. *Front. Nutr.* **2021**, *8*, 657090. [CrossRef]
8. Allai, F.M.; Azad, Z.A.A.; Mir, N.A.; Gul, K. Recent Advances in Non-Thermal Processing Technologies for Enhancing Shelf Life and Improving Food Safety. *Appl. Food Res.* **2023**, *3*, 100258. [CrossRef]
9. Chakka, A.K.; Sriraksha, M.; Ravishankar, C. Sustainability of Emerging Green Non-Thermal Technologies in the Food Industry with Food Safety Perspective: A Review. *LWT-Food Sci. Technol.* **2021**, *151*, 112140. [CrossRef]
10. Riekkinen, K.; Martikainen, K.; Korhonen, J. Effectiveness of High-Pressure Processing Treatment for Inactivation of Listeria Monocytogenes in Cold-Smoked and Warm-Smoked Rainbow Trout. *Appl. Sci.* **2023**, *13*, 3735. [CrossRef]
11. Khan, M.K.; Ahmad, K.; Hassan, S.; Imran, M.; Ahmad, N.; Xu, C. Effect of Novel Technologies on Polyphenols during Food Processing. *Innov. Food Sci. Emerg. Technol.* **2018**, *45*, 361–381. [CrossRef]
12. Pou, K.J.; Raghavan, V. Recent Advances in the Application of High Pressure Processing-Based Hurdle Approach for Enhancement of Food Safety and Quality. *J. Biosyst. Eng.* **2020**, *45*, 175–187. [CrossRef]
13. Velazquez, G.; Vázquez, P.; Vázquez, M.; Torres, J.A. Avances En El Procesado de Alimentos Por Alta Presión Advances in the Food Processing by High Pressure Avances No Procesado de Alimentos Por Alta Presión. *CYTA-J. Food.* **2005**, *4*, 353–367. [CrossRef]
14. Mújica-Paz, H.; Valdez-Fragoso, A.; Samson, C.T.; Welti-Chanes, J.; Torres, J.A. High-Pressure Processing Technologies for the Pasteurization and Sterilization of Foods. *Food Bioprocess Technol.* **2011**, *4*, 969–985. [CrossRef]
15. EFSA Panel on Biological Hazards (BIOHAZ Panel); Koutsoumanis, K.; Alvarez-Ordóñez, A.; Bolton, D.; Bover-Cid, S.; Chemaly, M.; Davies, R.; De Cesare, A.; Herman, L.; Hilbert, F. The Efficacy and Safety of High-pressure Processing of Food. *EFSA J.* **2022**, *20*, e07128.
16. Rendueles, E.; Omer, M.K.; Alvseike, O.; Alonso-Calleja, C.; Capita, R.; Prieto, M. Microbiological Food Safety Assessment of High Hydrostatic Pressure Processing: A Review. *LWT-Food Sci. Technol.* **2011**, *44*, 1251–1260. [CrossRef]
17. Wiśniewski, P.; Chajęcka-Wierzchowska, W.; Zadernowska, A. Impact of High-Pressure Processing (HPP) on Listeria Monocytogenes—An Overview of Challenges and Responses. *Foods* **2023**, *13*, 14. [CrossRef]
18. Milani, E.; Silva, F.V. Comparing High Pressure Thermal Processing and Thermosonication with Thermal Processing for the Inactivation of Bacteria, Moulds, and Yeasts Spores in Foods. *J. Food Eng.* **2017**, *214*, 90–96.
19. Morales, P.; Calzada, J.; Rodríguez, B.; De Paz, M.; Gaya, P.; Nuñez, M. Effect of Cheese Water Activity and Carbohydrate Content on the Barotolerance of Listeria Monocytogenes Scott A. *J. Food Prot.* **2006**, *69*, 1328–1333. [CrossRef]
20. Beuchat, L.R.; Komitopoulou, E.; Beckers, H.; Betts, R.P.; Bourdichon, F.; Fanning, S.; Joosten, H.M.; Ter Kuile, B.H. Low--Water Activity Foods: Increased Concern as Vehicles of Foodborne Pathogens. *J. Food Prot.* **2013**, *76*, 150–172. [CrossRef]
21. Medina-Meza, I.G.; Barnaba, C.; Barbosa-Cánovas, G.V. Effects of High Pressure Processing on Lipid Oxidation: A Review. *Innov. Food Sci. Emerg. Technol.* **2014**, *22*, 1–10. [CrossRef]
22. Guyon, C.; Meynier, A.; de Lamballerie, M. Protein and Lipid Oxidation in Meat: A Review with Emphasis on High-Pressure Treatments. *Trends Food Sci. Technol.* **2016**, *50*, 131–143. [CrossRef]
23. Andrés, A.I.; Cava, R.; Ventanas, J.; Muriel, E.; Ruiz, J. Lipid Oxidative Changes throughout the Ripening of Dry-Cured Iberian Hams with Different Salt Contents and Processing Conditions. *Food Chem.* **2004**, *84*, 375–381. [CrossRef]
24. Wang, Q.; Zhao, X.; Ren, Y.; Fan, E.; Chang, H.; Wu, H. Effects of High Pressure Treatment and Temperature on Lipid Oxidation and Fatty Acid Composition of Yak (*Poephagus grunniens*) Body Fat. *Meat Sci.* **2013**, *94*, 489–494. [CrossRef]
25. Cheah, P.; Ledward, D. Catalytic Mechanism of Lipid Oxidation Following High Pressure Treatment in Pork Fat and Meat. *J. Food Sci.* **1997**, *62*, 1135–1139. [CrossRef]
26. Aubourg, S.P.; Tabilo-Munizaga, G.; Reyes, J.E.; Rodríguez, A.; Pérez-Won, M. Effect of High-pressure Treatment on Microbial Activity and Lipid Oxidation in Chilled Coho Salmon. *Eur. J. Lipid Sci. Technol.* **2010**, *112*, 362–372. [CrossRef]
27. Lakshmanan, R.; Miskin, D.; Piggott, J.R. Quality of Vacuum Packed Cold-smoked Salmon during Refrigerated Storage as Affected by High-pressure Processing. *J. Sci. Food Agric.* **2005**, *85*, 655–661. [CrossRef]
28. Ma, H.; Ledward, D.; Zamri, A.; Frazier, R.; Zhou, G. Effects of High Pressure/Thermal Treatment on Lipid Oxidation in Beef and Chicken Muscle. *Food Chem.* **2007**, *104*, 1575–1579. [CrossRef]
29. Butz, P.; Tauscher, B. Emerging Technologies: Chemical Aspects. *Food Res. Int.* **2002**, *35*, 279–284. [CrossRef]
30. Rasanayagam, V.; Balasubramaniam, V.; Ting, E.; Sizer, C.; Bush, C.; Anderson, C. Compression Heating of Selected Fatty Food Materials during High-pressure Processing. *J. Food Sci.* **2003**, *68*, 254–259. [CrossRef]
31. Amadei, G.; Ross, B.M. Quantification of Character-impacting Compounds in *Ocimum basilicum* and'Pesto Alla Genovese'with Selected Ion Flow Tube Mass Spectrometry. *Rapid Commun. Mass Spectrom.* **2012**, *26*, 219–225. [CrossRef] [PubMed]
32. Klimankova, E.; Holadová, K.; Hajšlová, J.; Čajka, T.; Poustka, J.; Koudela, M. Aroma Profiles of Five Basil (*Ocimum basilicum* L.) Cultivars Grown under Conventional and Organic Conditions. *Food Chem.* **2008**, *107*, 464–472. [CrossRef]
33. Stanojevic, L.P.; Marjanovic-Balaban, Z.R.; Kalaba, V.D.; Stanojevic, J.S.; Cvetkovic, D.J.; Cakic, M.D. Chemical Composition, Antioxidant and Antimicrobial Activity of Basil (*Ocimum basilicum* L.) Essential Oil. *J. Essent. Oil Bear. Plants* **2017**, *20*, 1557–1569. [CrossRef]
34. Zengin, H.; Baysal, A.H. Antibacterial and Antioxidant Activity of Essential Oil Terpenes against Pathogenic and Spoilage-Forming Bacteria and Cell Structure-Activity Relationships Evaluated by SEM Microscopy. *Molecules* **2014**, *19*, 17773–17798. [CrossRef]

35. Arshad, H.M.; Mohiuddin, O.A.; Azmi, M.B. Comparative in Vitro Antibacterial Analysis of Different Brands of Cefixime against Clinical Isolates of Staphylococcus Aureus and *Escherichia coli*. *J. Appl. Pharm. Sci.* **2012**, *2*, 109–113.
36. Usaga, J.; Acosta, Ó.; Churey, J.J.; Padilla-Zakour, O.I.; Worobo, R.W. Evaluation of High Pressure Processing (HPP) Inactivation of *Escherichia coli* O157: H7, Salmonella Enterica, and Listeria Monocytogenes in Acid and Acidified Juices and Beverages. *Int. J. Food Microbiol.* **2021**, *339*, 109034. [CrossRef] [PubMed]
37. Ziaee, E.; Razmjooei, M.; Shad, E.; Eskandari, M.H. Antibacterial Mechanisms of Zataria Multiflora Boiss. Essential Oil against *Lactobacillus curvatus*. *LWT-Food Sci. Technol.* **2018**, *87*, 406–412. [CrossRef]
38. Klug, T.V.; Collado, E.; Martínez-Sánchez, A.; Gómez, P.A.; Aguayo, E.; Otón, M.; Artés, F.; Artés-Hernandez, F. Innovative Quality Improvement by Continuous Microwave Processing of a Faba Beans Pesto Sauce. *Food Bioprocess Technol.* **2018**, *11*, 561–571. [CrossRef]
39. Hilma, R.; Herliani, H.; Almurdani, M. Determination of Total Phenolic, Flavonoid Content Andfree Radical Scavenging Activity of Etanol Extract Sawo Stem Bark (*Manilkara zapota* (L.)). *Pros. CELSciTech* **2018**, *3*, 62–68.
40. Kulkarni, A.P.; Aradhya, S.M. Chemical Changes and Antioxidant Activity in Pomegranate Arils during Fruit Development. *Food Chem.* **2005**, *93*, 319–324. [CrossRef]
41. Bekdeşer, B.; Çelik, S.E.; Bener, M.; Dondurmacıoğlu, F.; Yıldırım, E.; Yavuz, E.N.; Apak, R. Determination of Primary and Secondary Oxidation Products in Vegetable Oils with Gold Nanoparticle Based Fluorometric Turn-on Nanosensor: A New Total Oxidation Value. *Food Chem.* **2024**, *434*, 137426. [CrossRef] [PubMed]
42. Hortwitz, W. AOAC Official Method 965.33, Peroxide Value of Oils and Fats. In *Official Methods of Analysis of AOAC International*, 17th ed.; AOAC Int.: Gaithersburg, MD, USA, 2002.
43. Tompkins, C.; Perkins, E.G. The Evaluation of Frying Oils with the P-anisidine Value. *J. Am. Oil Chem. Soc.* **1999**, *76*, 945–947. [CrossRef]
44. Yang, R.; Xu, J.; Lombardo, S.P.; Ganjyal, G.M.; Tang, J. Desiccation in Oil Protects Bacteria in Thermal Processing. *Food Res. Int.* **2020**, *137*, 109519. [CrossRef] [PubMed]
45. Yang, R.; Xie, Y.; Lombardo, S.P.; Tang, J. Oil Protects Bacteria from Humid Heat in Thermal Processing. *Food Control* **2021**, *123*, 107690. [CrossRef]
46. Georget, E.; Sevenich, R.; Reineke, K.; Mathys, A.; Heinz, V.; Callanan, M.; Rauh, C.; Knorr, D. Inactivation of Microorganisms by High Isostatic Pressure Processing in Complex Matrices: A Review. *Innov. Food Sci. Emerg. Technol.* **2015**, *27*, 1–14. [CrossRef]
47. D'Souza, T.; Karwe, M.; Schaffner, D.W. Effect of High Hydrostatic Pressure on Salmonella Inoculated into Creamy Peanut Butter with Modified Composition. *J. Food Prot.* **2014**, *77*, 1664–1668. [CrossRef]
48. Raffalli, J.; Rosec, J.; Carlez, A.; Dumay, E.; Richard, N.; Cheftel, J. High Pressure Stress and Inactivation of Listeria Innocua in Inoculated Dairy Cream. *Sci. Aliments* **1994**, *14*, 349–358.
49. Evert-Arriagada, K.; Trujillo, A.; Amador-Espejo, G.; Hernández-Herrero, M. High Pressure Processing Effect on Different *Listeria Spp.* in a Commercial Starter-Free Fresh Cheese. *Food Microbiol.* **2018**, *76*, 481–486. [CrossRef]
50. Romano, R.; De Luca, L.; Aiello, A.; Pagano, R.; Di Pierro, P.; Pizzolongo, F.; Masi, P. Basil (*Ocimum basilicum* L.) Leaves as a Source of Bioactive Compounds. *Foods* **2022**, *11*, 3212. [CrossRef]
51. Tarchoune, I.; Sgherri, C.; Izzo, R.; Lachaâl, M.; Navari-Izzo, F.; Ouerghi, Z. Changes in the Antioxidative Systems of *Ocimum basilicum* L. (Cv. Fine) under Different Sodium Salts. *Acta Physiol. Plant.* **2012**, *34*, 1873–1881. [CrossRef]
52. Şayin Sert, T.; Coşkun, F. The Effects of High-Pressure Processing on pH, Thiobarbituric Acid Value, Color and Texture Properties of Frozen and Unfrozen Beef Mince. *Molecules* **2022**, *27*, 3974. [CrossRef] [PubMed]
53. Okpala, C.O.; Piggott, J.R.; Schaschke, C.J. Influence of High-Pressure Processing (HPP) on Physico-Chemical Properties of Fresh Cheese. *Innov. Food Sci. Emerg. Technol.* **2010**, *11*, 61–67. [CrossRef]
54. Ciriello, M.; Formisano, L.; El-Nakhel, C.; Kyriacou, M.C.; Soteriou, G.A.; Pizzolongo, F.; Romano, R.; De Pascale, S.; Rouphael, Y. Genotype and Successive Harvests Interaction Affects Phenolic Acids and Aroma Profile of Genovese Basil for Pesto Sauce Production. *Foods* **2021**, *10*, 278. [CrossRef] [PubMed]
55. Kaşikçi, M.B.; Bağdatlioğlu, N. High Hydrostatic Pressure Treatment of Fruit, Fruit Products and Fruit Juices: A Review on Phenolic Compounds. *Food Health* **2016**, *2*, 27–39.

applied sciences

MDPI

Article

Characterization of Oxygenated Propenylbenzene Derivatives Binding to MAO-A Using Isothermal Titration Calorimetry and Molecular Modeling

Joanna Grzelczyk [1,*], Horacio Pérez-Sánchez [2], Miguel Carmena-Bargueño [2], Ilona Gałązka-Czarnecka [1], Grażyna Budryn [1], Dawid Hernik [3], Elisabetta Brenna [4] and Filip Boratyński [3]

1 Institute of Food Technology and Analysis, Faculty of Biotechnology and Food Sciences, Lodz University of Technology, 90-537 Lodz, Poland; ilona.galazka-czarnecka@p.lodz.pl (I.G.-C.); grazyna.budryn@p.lodz.pl (G.B.)

2 Structural Bioinformatics and High-Performance Computing Research Group (BIO-HPC), Computer Engineering Department, Universidad Católica de Murcia (UCAM), Guadalupe, 30107 Murcia, Spain; hperez@ucam.edu (H.P.-S.); mcarmena@ucam.edu (M.C.-B.)

3 Department of Food Chemistry and Biocatalysis, Wroclaw University of Environmental and Life Sciences, 50-375 Wroclaw, Poland; dawid.hernik@upwr.edu.pl (D.H.); filip.boratynski@upwr.edu.pl (F.B.)

4 Dipartimento di Chimica, Materiali ed Ingegneria Chimica "Giulio Natta", Politecnico di Milano, I-20133 Milan, Italy; mariaelisabetta.brenna@polimi.it

* Correspondence: joanna.grzelczyk@p.lodz.pl

Abstract: Monoamine oxidase A (MAO-A) is the main enzyme that deaminates serotonin. Correct serotonin concentration regulates appetite, improves well-being, reduces symptoms of depression, but also improves memory and heart function. In this study, the use of new compounds chemo-enzymatically synthesized from propenylbenzene derivatives as MAO-A inhibitors was proposed. Isothermal titration calorimetry analysis and molecular modeling were used to determine the inhibitory effect of MAO-A. The main compounds such as propenylbenzenes were observed to have no inhibitory effect. However, in the case of diols and hydroxy ketones, the compounds showed MAO-A inhibitory effects. In particular, 1-(4-hydroxy-3-methoxyphenyl)propane-1,2-diol, both in isothermal titration calorimetry (ITC) and molecular modeling, showed high affinity for the MAO-A enzyme, thus protecting serotonin from deamination. The results suggest that propenylbenzene derivatives, such as diols and hydroxy ketones, bind to MAO-A at the active site. This suggests their potential to be used as drugs or food supplements to prevent depression. These studies may constitute a new tool for further in vivo and in vitro studies.

Keywords: MAO-A; serotonin; docking simulation; propenylbenzenes; isothermal titration calorimetry

1. Introduction

5-hydroxytryptamine (5-HT) is a biogenic monoamine that performs various functions in the central nervous system (CNS). 5-HT is formed by the hydrolysis of tryptophan catalyzed by tryptophan hydrolase to 5-hydroxytryptophan and then decarboxylation to 5-HT [1]. 5-HT is synthesized and stored in the catecholaminergic, serotonergic, and pineal glands. If 5-HT synthesis occurs outside the CNS, it will occur in enterochromaffin cells and may be produced and stored in platelets to a small extent [1,2].

The incorrect level of 5-HT in the human body affects the pathogenesis of diseases such as irritable bowel syndrome, depression, appetite disorders, migraine, hypertension, and neurodegenerative diseases, and prevents diabetes [3–8]. Regulated 5-HT levels lead to better well-being, daily activity, and affective and cognitive functions. Therefore, serotonin reuptake inhibitors are often used in antidepressants for obsessive–compulsive disorder. Unfortunately, although they are effective drugs, they have several side effects [9].

Appropriate levels of 5-HT reduce obesity by improving fatty acid oxidation and regulating 5-HT signaling pathways. An increased level of the signaling pathway involved in combining 5-HT and MAO-A most often causes excessive liver serotonin degradation. This affects the overproduction of reactive oxygen species in liver mitochondria, which leads to obesity [10–13]. Other studies indicate that 5-HT has a vasoconstrictor or vasodilator function, depending on the binding of 5-HT to bradycardia or tachycardia. Serotonin is often administered as a pharmacological agent, as a therapeutic compound in the treatment of heart disease, hypertension, and migraine [14–18]. Binding serotonin to the 5-HT4 receptor affects the proper contraction of the heart muscle, and too high 5-HT concentration may lead to ischemia and cardiac arrhythmias [19–21].

Monoamine oxidoreductases (MAOs) are amine oxidases that break neurotransmitters such as dopamine, 5-HT, epinephrine, and noradrenaline [22,23]. MAO-A is the dominant form of MAOs that metabolizes 5-HT. Therefore, it plays a key role in the treatment of depression at every stage of the disease, chronic low mood, but also emotional disorders and obsessive–compulsive behaviors [24,25]. In chronic depressive disorder, increased levels of MAO-A and decreased levels of neurotransmitters are identified in the brain [26]. MAO-A increases appetite and contributes to the accumulation of adipocyte lipids and the differentiation of preadipocytes [27]. Ligands binding to the active site of MAO-A increase serotonic neurotransmission by inhibiting MAO-A. The MAO-A inhibitors used may cause adverse effects such as drowsiness. They may interact with food ingredients and drugs [28]. Therefore, it is important to look for new, non-toxic MAO-A inhibitors.

In this study, propenylbenzene derivatives such as diols and hydroxy ketones were used as candidates for MAO-A inhibitors. These compounds are used in various industries, e.g., in the food industry as food flavors and in cosmetics as essential oils. However, obtaining these compounds is not environmentally friendly. Therefore, an environmentally friendly chemo-enzymatic synthesis method was used, followed by whole-cell transformation, to obtain the studied compounds [29]. This study aimed to evaluate the degree of inhibition of MAO-A by propenylbenzene derivatives using molecular docking and isothermal titration calorimetry (ITC) methods. The innovation of this work is the first study of the enzymatic activity of MAO-A using a new chemo-enzymatic method of obtaining propenylbenzene derivatives.

2. Materials and Methods

2.1. Materials and Reagents

Prop-1-en-1-yl benzene (≥97%), isosafrole (≥98%), anethole (≥98%), 1,2-dimethoxy-4-prop-1-en-1-yl benzene (≥97%), isoeugenol (≥98%), Monoamine Oxidase A human (recombinant, expressed in baculovirus infected BTI insect cells, ≥99%), and serotonin (≥95%) were purchased from Sigma-Aldrich Chemical Co. (St. Louis, MO, USA).

The chemo-enzymatic synthesis of diols and hydroxy ketones has been described in a previous paper [29]. The following compounds were used for this study: isosafrole (**1a**), prop-1-en-1-yl benzene (**2a**), anethole (**3a**), 1,2-dimethoxy-4-prop-1-en-1-yl benzene (**4a**), isoeugenol (**5a**), 1-(1,3-benzodioxol-5-yl)propane-1,2-diol (**1b**), 1-phenylpropane-1,2-diol (**2b**), 1-(4-methoxyphenyl)propane-1,2-diol (**3b**),1-(3,4-dimethoxyphenyl)propane-1,2-diol (**4b**), 1-(4-hydroxy-3-methoxyphenyl)propane-1,2-diol (**5b**), 2-hydroxy-1-phenylpropan-1-one (**2c**), 2-hydroxy-1-(4-methoxyphenyl)propan-1-one (**3c**), and 1-(3,4-dimethoxyphenyl)-2-hydroxypropan-1-one (**4c**). The structure of the compounds is shown in Figure 1.

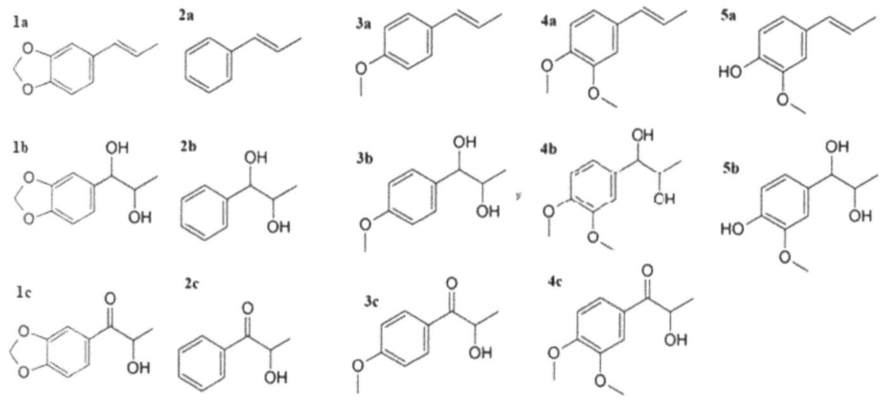

Figure 1. Structures of the propenylbenzenes: **1a–5a**; diols: **1b–5b**; hydroxy ketones: **1c–4c**.

2.2. Molecular Docking

The three-dimensional X-ray crystal structure from human MAO-A was downloaded from the protein data bank (http://www.rcsb.org Accessed on 5 July 2023) with PDB code 2BXR. The structural modeling involved three steps. First, a full atom model of the enzyme MAO-A was prepared from the PDB structure by calculating the ionizable amino acids' pKa values and protonation states. Next, hydrogen atoms were added. Finally, the cap termini were included with the Protein Preparation Wizard (Maestro, software version 2020-4). The partial charges of the protein atoms were added using System Builder (Maestro). The chemical structures of propenylbenzene derivatives were built up manually. Ligand partial charges were calculated as previously reported [30] for docking simulations. The docking of all compounds to the enzyme structural model was performed using the default configuration parameters of both docking software AutoDock Vina (AD) (software 1.1.2) and Lead Finder (LF) (v.2.0) using the MetaScreen software suite (https://github.com/bio-hpc/metascreener, Accessed on 5 July 2023). The protein and ligand files for docking were obtained in pdbqt format for AD and in mol2 format for LF. Two-dimensional diagrams for the obtained docking poses were prepared using the Maestro tool Ligand Interaction Diagram.

2.3. Isothermal Titration Calorimetry

The ITC MicroCal PEAQ-ITC200 apparatus (Malvern, Worcestershire, UK) was used for the calorimetric titration at 36.6 °C, as described previously [30]. The treatments included 19 injections, with 150 s between injections. A stirring speed of 250 rpm and 10 µcal/s of reference power were used. A measuring cell of 0.2 mL was filled with degassed MAO-A solution, 100 µmol/L, and titrated with each of the propenylbenzene derivatives at a concentration of 10 mmol/L or/and 5-HT (10 mmol/L). The enzyme and compounds were prepared in methanol LCMS, pH 7.6. Control experiments were performed by injecting MAO-A into the buffer in an identical manner and the resulting heat changes were subtracted from the measured heats of binding. The binding and thermodynamics constants were determined using the "One Set of Sites" fitting model. Stoichiometry (n), enthalpy changes (ΔH), Gibbs free energy changes (ΔG), entropy changes (ΔS), the dissociation constant (K_D), the binding constant (K_A), and the inhibition constant (K_i) were calculated from the ITC titration nonlinear least squares saturation curve fitting carried out using MicroCal PEAQ-ITC200 software 1.30. The K_i values for the oxidation of 5-HT catalyzed by MAO-A were calculated by using the Michaelis–Menten equation (Figure 2). To assess the inhibition activity, the signals of 5-HT oxidation in the presence of inhibitors were subtracted from the signals of 5-HT oxidation without inhibitors, while considering the enthalpy recorded during the interactions of the inhibitors with MAO-A.

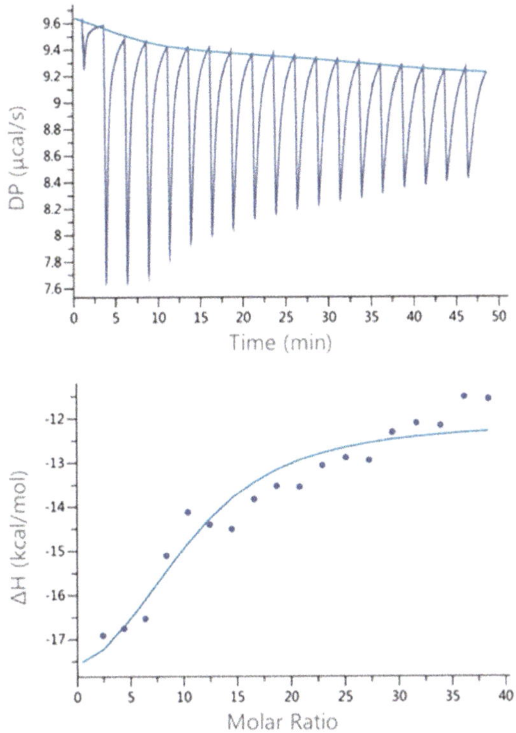

Figure 2. Example ITC raw data graph. The curve in the bottom figure shows the fitting of data to a one-set model, the nonlinear Levenberg–Marquardt fitting algorithm. The curve in the top figure is the baseline.

The inhibition activity was calculated as $[(\Delta H - \Delta H_i)/\Delta H] \times 100\%$, where ΔH_i is the signal of 5-HT oxidation in the presence of an inhibitor, and ΔH is the signal of 5-HT oxidation without an inhibitor. The percentage of inhibition of 5-HT oxidation at a given concentration of the ligand was calculated and it was used to calculate the concentration of ligands that inhibited 50% of the oxidation reaction (IC_{50}) [30]. The curves are provided in the supplement (Figure S1).

2.4. *Statistical Analysis*

The results were expressed as mean $\pm$ standard error of the mean, n = 3. All calculations were evaluated for significance using one-way ANOVA followed by Dunnett's test with the GraphPad Prism 6.0 software (GraphPad Software, Inc., La Jolla, CA, USA). Significance was defined at $p \leq 0.05$.

3. Results

3.1. *ITC Analysis*

ITC analysis was performed to determine the binding parameters of MAO-A with propenylbenzene derivatives. Studies have shown that for MAO-A, only compounds **1b**, **2b**, **2c**, **3b**, **3c**, **4b**, and **5b** bind to the enzyme in the active site, blocking the deamination of 5-HT. Table 1 shows the results of the thermodynamic parameters obtained from the analysis.

Table 1. Thermodynamic parameters of interactions between MAO-A and propenylbenzene derivatives.

Compound	K_D (mmol/L)	$K_A \times 10^3$ (L/mol)	ΔH (kJ/mol)	ΔG (kJ/mol)	ΔS (J/mol·K)	Inhibitor Activity	IC_{50} (µmol/L Inhibitor: µmol/L MAO-A)	K_i (µmol/L) K_M 5-HT 0.35 mmol/L
1a	nd	nd	nd	nd	nd	nd	nd	nd
1b	0.08 ± 0.00 a	12.19 ± 0.15 a	−38.18 ± 1.55 a	−34.00 ± 1.49 a	−0.01 ± 0.00 a	89.57 ± 1.15 a	3.54 ± 0.45 a	12.60 ± 0.02 a
1c	nd	nd	nd	nd	nd	nd	nd	nd
2a	nd	nd	nd	nd	nd	nd	nd	nd
2b	0.24 ± 0.01 b	4.18 ± 0.45 b	−2.00 ± 0.15 b	−21.48 ± 0.95 b	−0.06 ± 0.00 b	88.33 ± 1.29 a	4.76 ± 0.15 a	13.00 ± 0.01 a
2c	0.98 ± 0.03 c	1.02 ± 0.45 d	−27.21 ± 0.10 d	−17.88 ± 0.59 c	−0.03 ± 0.00 a	71.81 ± 1.82 b	6.33 ± 0.45 b	12.72 ± 0.02 a
3a	nd	nd	nd	nd	nd	nd	nd	nd
3b	0.02 ± 0.00 a	53.60 ± 2.03 c	−38.44 ± 1.05 a	−34.12 ± 0.55 a	−0.01 ± 0.00 a	95.22 ± 1.05 a	3.02 ± 0.10 a	<0.001
3c	0.25 ± 0.01 b	4.03 ± 0.05 b	−2.91 ± 0.59 b	−21.39 ± 1.08 b	0.06 ± 0.00 b	92.58 ± 1.10 a	3.25 ± 0.05 a	12.91 ± 0.02 a
4a	nd	nd	nd	nd	nd	nd	nd	nd
4b	0.25 ± 0.02 b	4.07 ± 0.39 b	−4.13 ± 0.55 c	−21.44 ± 0.95 b	0.06 ± 0.00 b	79.50 ± 0.35 b	5.72 ± 0.39 b	0.65 ± 0.01 b
4c	nd	nd	nd	nd	nd	nd	nd	nd
5a	nd	nd	nd	nd	nd	nd	nd	nd
5b	0.08 ± 0.01 a	12.21 ± 1.15 a	−40.53 ± 1.25 e	−24.24 ± 0.75 b	−0.05 ± 0.00 b	96.71 ± 1.92 a	2.38 ± 0.55 c	<0.001

Values are expressed as mean value ± SD; n = 3; a–e: different letters in one column or no index corresponds to significant differences ($p < 0.05$); nd: not detected.

The binding constant (K_A) of propenylbenzene derivatives ranged from 1.02 to 53.60×10^3 L/mol for **2c** and **3b**, respectively. Compounds **2b**, **3c**, and 4b showed a similar K_A of approximately 4×10^3 L/mol. However, compounds **1b** and **5b** showed a three times higher KA value than **2b**, **3c**, and **4b**. All tested compounds showed a negative enthalpy reaction (ΔH), indicating the exothermic effect of the reaction. The ΔH range was from −40.53 to −2.00 kJ/mol for **5b** and **2b**, respectively. Compounds **1b** and **3b** showed a slightly higher ΔH, approximately −38 kJ/mol, followed by **2c** (−27.21 kJ/mol). The highest negative free energy of −34.12 kJ/mol (ΔG) and −34.00 kJ/mol, interpreted as the substances affinity for binding, was recorded for **3b** and **1b**, while the lowest for **2c** was −17.88 kJ/mol. However, **2b**, **3c**, and **4b** showed a similar ΔG of approximately −21 kJ/mol. Negative reaction entropy (ΔS) was observed for compounds **1a**, **2b**, **2c**, **3b**, and **5b**, and positive for **3c** and **4b**.

The inhibition constant K_i of propenylbenzene derivatives indicated that the compounds bound to MAO-A as competitive inhibitors. Compounds **3b** and **5b** had aligned and low K_i, <0.001 µmol/L. The rest of the interactions showed values in the range of 0.65–13.00 µmol/L, respectively, for **4b** and **2b** (Table 1). The study showed that **5b** showed high inhibiting activity with IC_{50} amounting to 2.38 µmol/µmol of the enzyme. Compounds **1b**, **3b**, and **3c** showed slightly higher IC_{50} value, amounting to about 3 µmol/µmol of the enzyme. The study by Tao et al. showed that eugenol is a substance that protects against the cytotoxic effects of amyloid β peptides and has an antidepressant effect. It has been demonstrated that eugenol competitively inhibits MAO-A at K_i = 26 µmol/L; however, when testing the available structures related to eugenol, not every compound showed MAO-A inhibitory effects [31]. Our research reached a similar conclusion about isoeugenol (**5a**) and eugenol itself did not show any inhibitory effect; however, the already synthesized diol **5b** was the best MAO-A inhibitor, also of competitive type (K_i < 0.001).

3.2. Molecular Docking Studies

Two molecular docking techniques were used to characterize the binding of MAO-A to propenylbenzene derivatives: LeadFinder (LF) and AutoDock (AD). Each ligand and the protein were formatted separately, and calculations were performed. LF is a semi-empirical technique with high accuracy for generating ligand conformation algorithms. The results we obtain in this method are the binding energy and the predicted docking position of the ligand to the protein [32]. In contrast, AD exhibits ligand–protein integration, which exhibits larger spatial conformation surfaces and an accurate assessment of binding energies, which are most often used in drug design and discovery [33]. Modeling was performed for substances that showed inhibitory properties for MAO-A, protecting serotonin against deamination. The results of both analyses are presented in Figures 3 and 4 and Table 2.

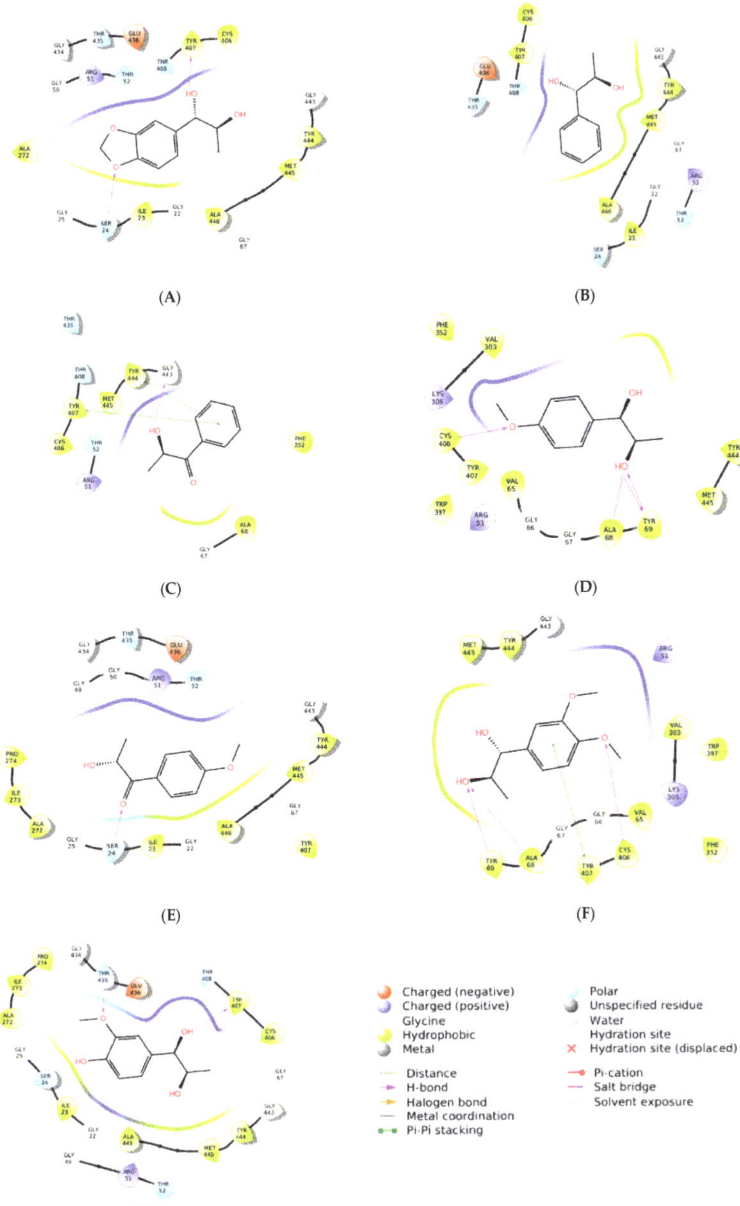

Figure 3. Two-dimensional representation of interaction diagrams of propenylbenzene derivative–MAO-A interactions in the active sites of the protein, AD. **(A) 1b**, **(B) 2b**, **(C) 2c**, **(D) 3b**, **(E) 3c**, **(F) 4b**, **(G) 5b**.

The analysis carried out allowed us to determine the affinity of MAO-A for propenylbenzene derivatives precisely. According to AD and LF docking, **1b** showed the highest affinity for actin, with ΔG amounting to −31.18 kJ/mol (AD) and −30.64 kJ/mol (LF), and **5b** showed high ΔG, amounting to −27.21 kJ/mol (AD) and −32.94 kJ/mol (LF). In the LF molecular modeling technique, the tested compounds interacted through H-bonds and π-cations. The highest number of H-bonds occurred for **5b** (Figure 4G), linking to the amino acids residues of GLY443, TYR407, THR435, and SER23, and for **1b** (Figure 4A) to TYR407, GLY443, THR435, and SER24. Compounds **3b**, **3c**, and **4b** (Figure 4D–F) bound to MAO-A in the active site with the residues of TYR407 and SER24. However, compounds **2b** and **2c**, in addition to H-bonds, bound to the enzyme in the active site via a π-cation (Figure 4B,C). Compound **2b** bound by H-bonds to the amino acid residues of ILE23, ARG51, and GLY49 and by π-cation to ARG51. Compound **2c**, in addition to the π-cation binding to the residue of ARG51, bound to THR435, ALA272, and AGR51.

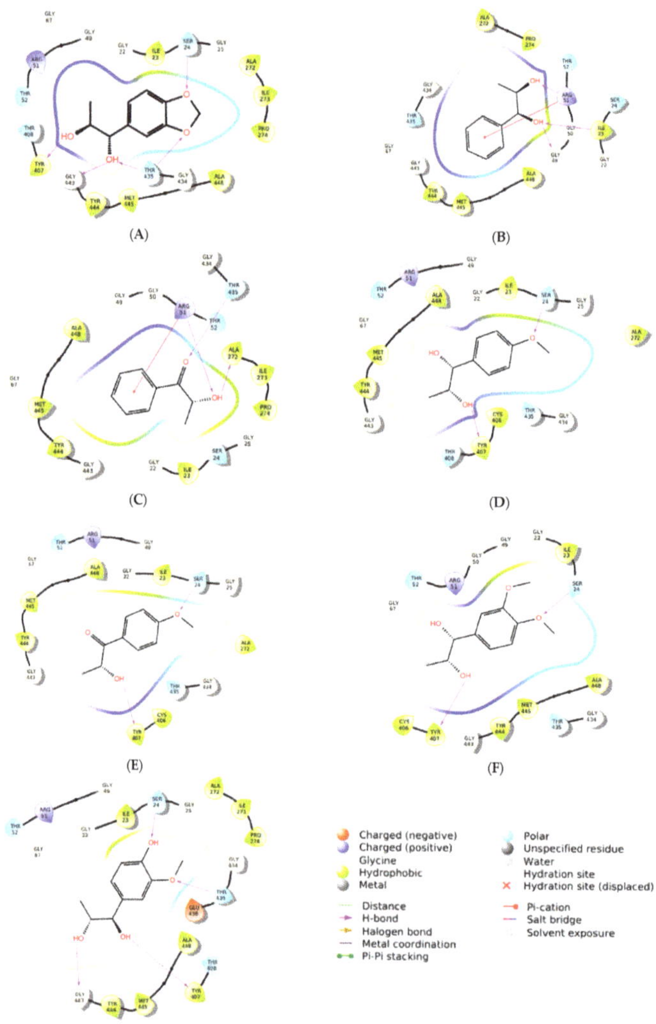

Figure 4. Two-dimensional representation of interaction diagrams of propenylbenzene derivative–MAO-A interactions in the active sites of the protein, LF. **(A) 1b**, **(B) 2b**, **(C) 2c**, **(D) 3b**, **(E) 3c**, **(F) 4b**, **(G) 5b**.

Table 2. Propenylbenzene derivative (diol and hydroxy ketone) interactions with identical amino acid residues on MAO-A.

Compound	1b	2b	2c	3b	3c	4b	5b
AD Docking Score (kJ/mol)	−31.18	−25.71	−26.63	−26.58	−27.25	−26.71	−27.21
LF Docking Score (kJ/mol)	−30.64	−26.28	−23.28	−26.96	−25.33	−29.72	−32.94
ILE23	AD and LF	AD and LF	No	No	AD and LF	LF	AD
SER24	AD and LF	No	No	LF	AD and LF	LF	AD and LF
GLY49	No	LF	No	No	No	No	No
ARG51	No	AD and LF	AD and LF	LF	AD and LF	LF	AD and LF
THR52	No	AD and LF	LF	LF	AD and LF	LF	LF
GLY67	No	No	No	LF	AD and LF	LF	No
ALA68	No	No	No	AD	No	AD	No
TYR69	No	No	No	AD	No	AD	No
ALA272	No	No	LF	No	No	No	No
VAL303	No	No	No	No	No	No	No
LYS305	No	No	No	No	No	No	No
PHE352	No	No	No	AD	No	AD	No
TYR407	AD and LF	AD	AD	AD and LF	LF	AD and LF	AD and LF
THR435	AD and LF	AD and LF	LF	LF	AD and LF	LF	AD and LF
GLY443	AD and LF	No	AD	No	No	No	LF
TYR444	No	No	AD	Ad	No	AD	AD
MET445	No	AD	AD	No	No	No	No
ALA448	AD and LF	AD and LF	LF	LF	AD and LF	LF	No

The results of molecular modeling using the AD technique were characterized by fewer H-bonds and the occurrence of π-π interactions. The π-π interactions occurred for compounds **2c** and **4b**. Compound **2c** is connected via a π-π interaction with TYR444 and TYR407 and by H-bond with GLY443 (Figure 3C). Compound **4b** bound by the π-π bond to the amino acid TYR47 and by H-bond with TYR69, ALA68, and CYS406 (Figure 3F). Compounds **1b**, **2b**, **3b**, **3c**, and **5b** were bound to the enzyme by H-bonds. Compound **1b** bound to MAO-A via an H-bond to the amino acid residues of SER24 and TYR407 (Figure 3A). Compound **3b** bound to MAO-A via an H-bond to CYS46, ALA68, and TYR69 (Figure 3D), and for **5b**, bonds with TYR407 and THR435 were calculated (Figure 3G). However, neither H-bond binding nor π-π interactions were recorded for compound **2b** (Figure 3B). Most of the compounds are bound with amino acid residues of TYR. Compounds hydrogen-bonded with TYR gave higher affinities for the enzyme.

4. Discussion

5-HT is a key neurotransmitter in the CNS, gastrointestinal tract, and platelets. It performs many important functions in the human body, such as the following: it regulates intestinal peristalsis; it is a vasoconstrictor and helps maintain homeostasis balance and contributes to proper blood clotting. The function of 5-HT in the CNS is not fully understood. So far, it is known that it is responsible for cognitive functions and regulates mood and appetite. Although there is only 10% of serotonin in the CNS, its reduction affects the occurrence of depression, obsessive–compulsive disorders, and anxiety states. Reduced 5-HT levels also contribute to developing heart disease and inflammation [34–36]. Therefore, proposing new, non-toxic compounds that inhibit MAO-A is essential. The studies documented that compounds such as diols and hydroxy ketones obtained using chemo-enzymatic methods from propenylbenzene bind to MAO-A in the active site. This means that they increase serotonin transport by inhibiting MAO-A. Additionally, by inhibiting deamination, the inactivation of MAO-A prevents the formation of intermediate products such as hydrogen peroxide and aldehydes, which contribute to inflammation through the formation of amyloid plaques and accompanying the degradation of neuronal membranes [34,37]. An ITC experiment was carried out to verify the possibility of binding the studied diols and hydroxy ketones in the active site of the enzyme under the physiological concentrations.

It was observed that compounds **1a**, **1c**, **2a**, **3a**, **4a**, **4c**, and **5a** did not inhibit MAO-A. The experiment was repeated at different larger concentrations to determine the relationship between concentration and inhibition activity. Interestingly, diols and hydroxy ketones were already bound in the active site of MAO-A, protecting serotonin against deamination. Such a relationship was observed for compounds **1b**, **2b**, **2c**, **3b**, **3c**, **4b**, and **5b**. The most stable bonds with MAO-A were formed by compounds **5b** ($\Delta H = -40.53$ kJ/mol; $\Delta G = -24.24$ kJ/mol; IC50 = 2.38 μmol/L) and **3b** ($\Delta H = -38.44$ kJ/mol; $\Delta G = -34.12$ kJ/mol; IC_{50} = 3.02 μmol/L). Our research found isoeugenol (**5a**) as an efficient MAO-A inhibitor. Eugenol did not show any inhibitory effect; however, the already synthesized diol **5b** was the best MAO-A inhibitor ($K_i < 0.001$). Drukarch's research noted that anethole dithiolethione, often used in treating neurodegenerative diseases, has an inhibitory effect on MAO-B, while it has no inhibitory effect on MAO-A [38]. Vastegani and co-authors also indicated the effect of anethole on the MAO-B enzyme [39]. This was confirmed in our studies, because anethol itself (**3a**) did not show MAO-A inhibitory properties, while the corresponding diol (**3b**) and hydroxy ketone (**3c**) showed MAO-A inhibition activity at an IC_{50} concentration of approx. 3 μmol/L in a competitive manner; for **3b**, K_i was <0.001 μmol/L, whereas for **3c**, it amounted to 12.91 μmol/L. The same relationship was observed for **4a**, but in this case, only compound **4b** showed an inhibitory effect with $K_i = 0.65$ μmol/L. For compounds **1a** and **2a**, no studies concerning MAO-A inhibition were found. However, it can be observed that both **1b** and **2b** interacted with MAO-A in the active site, while hydroxy ketones did so only in the case of compound **2b**. It can be concluded that the best family of MAO-A inhibitors tested in this work are the hydroxy ketones, which protect serotonin against deamination.

ITC is a technique that measures the heat of different reactions, not only binding. In the case of enzyme–substrate reactions, the device measured the heat released as a result of substrate binding to the enzyme. This can lead to a substrate reaction, e.g., its oxidation, as it was in the case of monoaminoxidase. The "Michaelis constant" mode was used to mathematically describe such a reaction. In the case of enzyme–inhibitor reactions alone, the device measured the heat of binding of the inhibitors to the enzyme. The "binding to one site" mode was used to describe the reaction. Importantly, in the case of enzyme–substrate reactions in the presence of an inhibitor, the "inhibition constant" mode was used. On that basis, IC50 was calculated. This showed that the ITC method is adjusted to characterize enzymatic reactions. The limitation of the ITC method is that the inhibitor may bind to the enzyme not only in the active site, giving an additional small heat release signal. For this reason, it is complementary to perform docking simulations, which allows us to describe the binding only in the active site. Therefore, molecular modeling was used to compare affinity of inhibitors to the enzyme [40–42].

It could be observed that both methods (ITC and molecular docking) showed similar trends for compounds **1b** and **5b** interacting with the enzyme. In both tests, compound **5b** was shown to be the best inhibitor of MAO-A, which binds to the active site competitively (Table 1). Molecular modeling was performed using two methods: AD and LF. In most cases, higher affinity for the enzyme was demonstrated by LF docking (Table 2). 5-HT binds to MAO-A through the amino acids ILE335 and TYR326 [43]. MAO-A forms covalent bonds with the substrate by cysteine residues, e.g., CYS406 [44]. An example of an effective inhibitor is compound **5b**, which showed high affinity for the enzyme in a docking procedure by binding to the enzyme through the amino acid residue TYR407 adjacent to CYS406, blocking access to the substrate.

Adding aromatic compounds such as propenylbenzene derivatives (hydroxy ketones and diols) to food can not only enrich it with an attractive aroma, but also be an additive, giving food functional properties. Preparations of aromatic compounds can also be used in pharmacy in the future. Both directions of potential application require many studies related to their safe use in the diet. The nearest planned studies will include adding the tested aromatic compounds to different types of food and assessing their bioavailability during simulated digestion and absorption in Caco-2 cells.

5. Conclusions

To sum up, these studies showed that compounds synthesized from propenylbenzene derivatives, namely, hydroxy ketones and diols, showed high binding affinity to MAO-A in the active sites, as shown by ITC and molecular docking methods. Hydroxy ketones and diols binding to MAO-A may increase and regulate the level of serotonin in the body, which may result in better well-being, reduced appetite, and improved cognitive functions. The most effective MAO-A inhibitor was 1-(4-hydroxy-3-methoxyphenyl)propane-1,2-diol, which, both in the ITC study and in molecular modeling, showed high affinity for the enzyme, forming a stable bond in the region of the active site through the TYR amino acid residue. The newly synthesized compounds can be functional food ingredients or used in pharmacy. The results of this study will be used in future research on the beneficial effect of the health-promoting properties of propenylbenzene derivatives, evaluated in complex in vitro and in vivo models.

Supplementary Materials: The following supporting information can be downloaded at: https://www.mdpi.com/article/10.3390/app142310856/s1, Figure S1: ITC raw data of interactions of MAO-A with (A) 1-(1,3-benzodioxol-5-yl)propane-1,2-diol, (C) 1-phenylpropane-1,2-diol, (E) 2-hydroxy-1-phenylpropan-1-one, (G) 2-hydroxy-1-(4-methoxyphenyl)propan-1-one, (I) 1-(4-methoxyphenyl)propane-1,2-diol, (K)1-(3,4-dimethoxyphenyl)propane-1,2-diol, (M) 1-(4-hydroxy-3-methoxyphenyl)propane-1,2-diol; Oxidation of 5-HT catalyzed by MAO-A at a presence of: (B) 1-(1,3-benzodioxol-5-yl)propane-1,2-diol, (D) 1-phenylpropane-1,2-diol, (F) 2-hydroxy-1-phenylpropan-1-one, (H) 2-hydroxy-1-(4-methoxyphenyl) propan-1-one, (J) 1-(4-methoxyphenyl)propane-1,2-diol, (L) 1-(3,4-dimethoxyphenyl)propane-1,2-diol, (N) 1-(4-hydroxy-3-methoxyphenyl)propane-1,2-diol. Curves in bottom figures show the fitting of data to a one-set model by the nonlinear Levenberg–Marquardt fitting algorithm supplied by the instrument.

Author Contributions: Conceptualization, J.G. and F.B.; methodology, J.G., H.P.-S., G.B., E.B. and F.B.; formal analysis, J.G., M.C.-B., I.G.-C. and D.H.; investigation, J.G., M.C.-B., I.G.-C. and D.H.; writing—original draft preparation, J.G.; writing—review and editing, H.P.-S., G.B., E.B. and F.B.; resources, D.H.; supervision, F.B.; funding acquisition, F.B. All authors have read and agreed to the published version of the manuscript.

Funding: This research received no external funding.

Institutional Review Board Statement: Not applicable.

Informed Consent Statement: Not applicable.

Data Availability Statement: The raw data supporting the conclusions of this article will be made available by the authors on request.

Acknowledgments: M.C.-B. is employed to the training of research staff financed by the Plan Propio de Investigación de la UCAM. Supercomputing infrastructure of the Extremadura Research Centre of Advanced Technologies (CETA-CIEMAT), funded by the European Regional Development Fund (ERDF). CETA-CIEMAT is part of CIEMAT and the Government of Spain.

Conflicts of Interest: The authors declare no conflicts of interest.

References

1. Mohammad-Zadeh, L.F.; Moses, L.; Gwaltney-Brant, S.M. Serotonin: A review. *J. Vet. Pharmacol. Ther.* **2008**, *31*, 187–199. [CrossRef] [PubMed]
2. Hernandez-Leon, A.; Escamilla-Orozco, R.I.; Tabal-Robles, A.R.; Martínez-Vargas, D.; Romero-Bautista, L.; Escamilla-Soto, G.; González-Romero, O.S.; Torres-Valencia, M.; González-Trujano, M.E. Antidepressant- and anxiolytic-like activities and acute toxicity evaluation of the Psilocybe cubensis mushroom in experimental models in mice. *J. Ethnopharmacol.* **2024**, *320*, 117415. [CrossRef] [PubMed]
3. Baganz, N.L.; Blakely, R.D. A Dialogue between the immune system and brain, spoken in the language of serotonin. *ACS ChemNeurosci.* **2013**, *4*, 48–63. [CrossRef]

4. Shajib, M.S.; Khan, W.I. The role of serotonin and its receptors in activation of immune responses and inflammation. *ACTA Physiol.* **2015**, *213*, 561–574. [CrossRef]
5. Deen, M.; Hansen, H.D.; Hougaard, A.; Nørgaard, M.; Eiberg, H.; Lehel, S.; Ashina, M.; Knudsen, G.M. High brain serotonin levels in migraine between attacks: A 5-HT4 receptor binding PET study. *Neuroimage Clin.* **2018**, *18*, 97–102. [CrossRef]
6. Aggarwal, M.; Puri, V.; Puri, S. Serotonin and CGRP in migraine. *Ann. Neurosci.* **2012**, *19*, 88–94. [CrossRef]
7. Zepf, F.D.; Dingerkus, V.L.; Helmbold, K.; Bubenzer-Busch, S.; Biskup, C.S.; Herpertz-Dahlmann, B.; Schaab, M.; Kratzsch, J.; Eisert, A.; Rink, L.; et al. Effects of a short-term reduction in brain serotonin synthesis on the availability of the soluble leptin receptor in healthy women. *J. Neural Transm.* **2015**, *122*, 343–348. [CrossRef]
8. Hua, X.; Zhu, J.; Yang, T.; Guo, M.; Li, Q.; Chen, J.; Li, T. The gut microbiota and associated metabolites are altered in sleep disorder of children with autism spectrum disorders. *Front. Psychiatry* **2020**, *11*, 855. [CrossRef]
9. Homberg, J.R. Serotonin and decision making processes. *Neurosci. Biobehav. Rev.* **2012**, *36*, 218–236. [CrossRef]
10. Li, H.Y.; Huang, S.Y.; Zhou, D.D.; Xiong, R.G.; Luo, M.; Saimaiti, A.; Han, M.K.; Gan, R.Y.; Zhu, H.L.; Li, H.B. Theabrownin inhibits obesity and non-alcoholic fatty liver disease in mice via serotonin-related signaling pathways and gut-liver axis. *J. Adv. Res.* **2023**, *52*, 59–72. [CrossRef]
11. Palamiuc, L.; Noble, T.; Witham, E.; Ratanpal, H.; Vaughan, M.; Srinivasan, S. A tachykinin-like neuroendocrine signalling axis couples central serotonin action and nutrient sensing with peripheral lipid metabolism. *Nat. Commun.* **2017**, *8*, 14237. [CrossRef] [PubMed]
12. Watanabe, H.; Rose, M.T.; Aso, H. Role of peripheral serotonin in glucose and lipid metabolism. *Curr. Opin. Lipidol.* **2011**, *22*, 186–191. [CrossRef] [PubMed]
13. El-Merahbi, R.; Löffler, M.; Mayer, A.; Sumara, G. The roles of peripheral serotonin in metabolic homeostasis. *FEBS Lett.* **2015**, *589*, 1728–1734. [CrossRef]
14. Nebigil, C.G.; Maroteaux, L. Functional consequence of serotonin/5-HT2B receptor signaling in heart: Role of mitochondria in transition between hypertrophy and heart failure? *Circulation* **2003**, *108*, 902–908. [CrossRef]
15. Noorlander, C.W.; Ververs, F.F.; Nikkels, P.G.; van Echteld, C.J.; Visser, G.H.; Smidt, M.P. Modulation of serotonin transporter function during fetal development causes dilated heart cardiomyopathy and lifelong behavioral abnormalities. *PLoS ONE* **2008**, *3*, e2782. [CrossRef]
16. Selim, A.M.; Sarswat, N.; Kelesidis, I.; Iqbal, M.; Chandra, R.; Zolty, R. Plasma serotonin in heart failure: Possible marker and potential treatment target. *Heart Lung Circ.* **2017**, *26*, 442–449. [CrossRef]
17. Bhuiyan, M.S.; Tagashira, H.; Fukunaga, K. Crucial interactions between selective serotonin uptake inhibitors and sigma-1 receptor in heart failure. *J. Pharmacol. Sci.* **2013**, *121*, 177–184. [CrossRef]
18. Panconesi, A. Serotonin and migraine: A reconsideration of the central theory. *J. Headache Pain* **2008**, *9*, 267–276. [CrossRef]
19. Neumann, J.; Hofmann, B.; Dhein, S.; Gergs, U. Cardiac roles of serotonin (5-HT) and 5-HT-receptors in health and disease. *Int. J. Mol. Sci.* **2023**, *24*, 4765. [CrossRef]
20. Nezafati, M.H.; Eshraghi, A.; Vojdanparast, M.; Abtahi, S.; Nezafati, P. Selective serotonin reuptake inhibitors and cardiovascular events: A systematic review. *J. Res. Med. Sci.* **2016**, *21*, 66. [CrossRef]
21. Levy, R.J. Serotonin transporter mechanisms and cardiac disease. *Circulation* **2006**, *113*, 2–4. [CrossRef] [PubMed]
22. Yang, Z.; Zhong, T.; Cao, B.; Liao, D.; Hu, X.; Zhao, S.; Qin, J. Monoamine oxidase A probe synthesis and fluorescence imaging in cells and zebrafish based on AIE mechanism. *Mater. Today Chem.* **2023**, *35*, 101890. [CrossRef]
23. Sarfraz, M.; Ibrahim, M.K.; Ejaz, S.A.; Attaullah, H.M.; Aziz, M.; Arafat, M.; Shamim, T.; Elhadi, M.; Ruby, T.; Mahmood, H.K. An integrated computational approaches for designing of potential piperidine based inhibitors of Alzheimer disease by targeting cholinesterase and monoamine oxidases isoenzymes. *Appl. Biochem. Biotechnol.* **2024**. [CrossRef] [PubMed]
24. Zhang, L.; Zhang, R.; Cong, X.; He, M.; Zhao, X.; Fan, J.; Peng, X.; Cui, J.; Sun, W. Near-infrared fluorescence probe for monoamine oxidase a with a large stokes shift for intraoperative navigation. *ACS Appl. Bio Mater.* **2024**, *7*, 1115–1124. [CrossRef]
25. Sharma, R.; Kumarasamy, M.; Parihar, V.K.; Ravichandiran, V.; Kumar, N. Monoamine oxidase: A potential link in Papez circuit to generalized anxiety disorders. *CNS Neurol. Disord. Drug Targets* **2023**, *23*, 638–655. [CrossRef]
26. Li, X.; Shi, D.; Song, Y.; Xu, Y.; Gao, Y.; Qiu, W.; Chen, X.; Li, X.; Huang, Y.; Feng, Y.; et al. Specific tracking of monoamine oxidase A in heart failure models by a far-red fluorescent probe with an ultra large Stokes shift. *Chin. Chem. Lett.* **2022**, *33*, 1572–1576. [CrossRef]
27. Niveta, J.P.S.; John, C.M.; Arockiasamy, S. Monoamine oxidase mediated oxidative stress: A potential molecular and biochemical crux in the pathogenesis of obesity. *Mol. Biol. Rep.* **2023**, *51*, 29. [CrossRef]
28. Goldberg, J.F.; Thase, M.E. Monoamine oxidase inhibitors revisited: What you should know. *J. Clin. Psychiatry* **2013**, *74*, 189–191. [CrossRef]
29. Hernik, D.; Szczepańska, E.; Ghezzi, M.C.; Brenna, E.; Włoch, A.; Pruchnik, H.; Mularczyk, M.; Marycz, K.; Olejniczak, T.; Boratyński, F. Chemo-enzymatic synthesis and biological activity evaluation of propenylbenzene derivatives. *Front. Microbiol.* **2023**, *14*, 1223123. [CrossRef]
30. Grzelczyk, J.; Budryn, G.; Peña-García, J.; Szwajgier, D.; Gałązka-Czarnecka, I.; Oracz, J.; Pérez-Sánchez, H. Evaluation of the inhibition of monoamine oxidase A by bioactive coffee compounds protecting serotonin degradation. *Food Chem.* **2021**, *348*, 129108. [CrossRef]

31. Tao, G.; Irie, Y.; Li, D.J.; Keung, W.M. Eugenol and its structural analogs inhibit monoamine oxidase A and exhibit antidepressant-like activity. *Bioorg. Med. Chem.* **2005**, *13*, 4777–4788. [CrossRef] [PubMed]
32. Nelen, J.; Carmena-Bargueño, M.; Martínez-Cortés, C.; Rodríguez-Martínez, A.; Villalgordo-Soto, J.M.; Pérez-Sánchez, H. ESSENCE-Dock: A consensus-based approach to enhance virtual screening enrichment in drug discovery. *ChemRxiv* **2023**. [CrossRef] [PubMed]
33. Goodsell, D.S.; Sanner, M.F.; Olson, A.J.; Forli, S. The AutoDock suite at 30. *Protein Sci.* **2021**, *30*, 31–43. [CrossRef] [PubMed]
34. Prah, A.; Purg, M.; Stare, J.; Vianello, R.; Mavri, J. How monoamine oxidase A decomposes serotonin: An empirical valence bond simulation of the reactive step. *J. Phys. Chem. B* **2020**, *124*, 8259–8265. [CrossRef] [PubMed]
35. Wu, H.; Denna, T.H.; Storkersen, J.N.; Gerriets, V.A. Beyond a neurotransmitter: The role of serotonin in inflammation and immunity. *Pharmacol. Res.* **2019**, *140*, 100–114. [CrossRef]
36. Naoi, M.; Maruyama, W.; Shamoto-Nagai, M. Type A monoamine oxidase and serotonin are coordinately involved in depressive disorders: From neurotransmitter imbalance to impaired neurogenesis. *J. Neural Transm.* **2018**, *125*, 53–66. [CrossRef]
37. Knittel, J.; Itani, N.; Schreckenberg, R.; Heger, J.; Rohrbach, S.; Schulz, R.; Schlüter, K.-D. Monoamine oxidase A contributes to serotonin—But not norepinephrine-dependent damage of rat ventricular myocytes. *Biomolecules* **2023**, *13*, 1013. [CrossRef]
38. Drukarch, B.; Flier, J.; Jongenelen, C.A.; Andringa, G.; Schoffelmeer, A.N. The antioxidant anethole dithiolethione inhibits monoamine oxidase-B but not monoamine oxidase A activity in extracts of cultured astrocytes. *J. Neural Transm.* **2006**, *113*, 593–598. [CrossRef]
39. Vastegani, M.S.; Khoshnam, S.E.; Ghafouri, S.; Bakhtiari, N.; Farbood, Y.; Sarkaki, A. Anethole attenuates motor dysfunctions, striatal neuronal activity deficiency and blood brain barrier permeability by decreasing striatal α-synuclein and oxidative stress in rotenone-induced Parkinson's disease of male rats. *PLoS ONE* **2023**, *18*, e0294612. [CrossRef]
40. Siddiqui, K.S.; Poljak, A.; Ertan, H.; Bridge, W. The use of isothermal titration calorimetry for the assay of enzyme activity: Application in higher education practical classes. *Biochem. Mol. Biol. Educ.* **2022**, *50*, 519–526. [CrossRef]
41. Wang, Y.; Wang, G.; Moitessier, N.; Mittermaier, A.K. Enzyme Kinetics by Isothermal Titration Calorimetry: Allostery, Inhibition, and Dynamics. *Front. Mol. Biosci.* **2020**, *7*, 583826. [CrossRef] [PubMed]
42. Abis, G.; Pacheco-Gómez, R.; Bui, T.T.T.; Conte, M.R. Isothermal Titration Calorimetry Enables Rapid Characterization of Enzyme Kinetics and Inhibition for the Human Soluble Epoxide Hydrolase. *Anal. Chem.* **2019**, *91*, 14865–14872. [CrossRef] [PubMed]
43. Jalal, K.; Khan, K.; Haleem, D.J.; Uddin, R. In silico study to identify new monoamine oxidase type a (MAO-A) selective inhibitors from natural source by virtual screening and molecular dynamics simulation. *J. Mol. Struct.* **2022**, *1254*, 132244. [CrossRef]
44. Kaur, A.; Garg, S.; Shiekh, B.A.; Singh, N.; Singh, P.; Bhatti, R. In silico studies and in vivo MAOA inhibitory activity of coumarins isolated from Angelica archangelica extract: An approach toward antidepressant activity. *ACS Omega* **2020**, *5*, 15069–15076. [CrossRef] [PubMed]

applied sciences

MDPI

Review

How Hydrogen (H_2) Can Support Food Security: From Farm to Fork

Grace Russell [1,*], Alexander Nenov [2] and John T. Hancock [1,*]

1 School of Applied Sciences, University of the West of England, Bristol BS16 1QY, UK
2 Water Fuel Engineering, Church Street, Wakefield WF1 5QY, UK; nenov.alexander@waterfuelengineering.com
* Correspondence: grace.russell@uwe.ac.uk (G.R.); john.hancock@uwe.ac.uk (J.T.H.)

Abstract: Molecular hydrogen (H_2) is a low-molecular-weight, non-polar and electrochemically neutral substance that acts as an effective antioxidant and cytoprotective agent, with research into the effects of H_2 incorporation into the food chain, at various stages, rapidly gaining momentum. H_2 can be delivered throughout the food growth, production, delivery and storage systems in numerous ways, including as a gas, as hydrogen-rich water (HRW), or with hydrogen-donating food supplements such as calcium (Ca) or magnesium (Mg). In plants, H_2 can be exploited as a seed-priming agent, during seed germination and planting, during the latter stages of plant development and reproduction, as a post-harvest treatment and as a food additive. Adding H_2 during plant growth and developmental stages is noted to improve the yield and quality of plant produce, through modulating antioxidant pathways and stimulating tolerance to such environmental stress factors as drought stress, enhanced tolerance to herbicides (paraquat), and increased salinity and metal toxicity. The benefits of pre- and post-harvest application of H_2 include reductions in natural senescence and microbial spoilage, which contribute to extending the shelf-life of animal products, fruits, grains and vegetables. This review collates empirical findings pertaining to the use of H_2 in the agri-food industry and evaluates the potential impact of this emerging technology.

Keywords: hydrogen; molecular hydrogen; food; food safety; non-thermal food preservation

1. Introduction

As the global population increases, so do the obligations to provide nutritious sustenance to an ageing population. For this to be achieved, the preservation of fresh produce is critical. Due to such contemporary issues as climate change, chemical pollution, loss of biodiversity and soil erosion, the production, preservation, and distribution of quality foods becomes ever more challenging [1]. Currently, such preservation processes as cooking, drying, pasteurization, refrigeration and sterilisation can make food products more resistant to microbial spoilage and attenuate lipid oxidation, thereby decelerating the natural decomposition and rancidification of produce over time [2]. Although effective, such techniques can be expensive to implement and maintain; there is, therefore, a requirement to develop alternative, ecologically responsible, low cost and sustainable methods of preserving natural produce.

Molecular hydrogen (H_2) is classified as a 'generally regarded as safe' (GRAS) product by the U.S. Food and Drug Administration [3] and is regarded as a food supplement (E949) under part C group I of regulation 1129/2011 in the European Union [4] and U.K [5]. H_2 technologies are at the forefront of modern research, due in part to it being trialled as an efficient and sustainable source of energy [6] and in part to its recognition as an effective antioxidant and cytoprotective compound in health and agricultural science [7,8]. In this regard, H_2 could provide the increased energy required for sustainable food production, storage and distribution while enhancing food security through maintaining the nutritional value and prolonging the shelf-life of produce.

Oxidative stress and subsequent cellular and tissue damage can be caused by numerous factors, including environmental challenges (e.g., drought, pollution, increased soil salinity), human intervention (e.g., fertilisers, insecticides, paraquat) and natural senescence (e.g., cellular deterioration, microbial spoilage). These factors can affect the growth, yield and nutritional content of food [8]. Furthermore, oxidative damage can lead to the rapid deterioration of food quality, resulting in lipid oxidation, discolouration and alterations to both the nutritional profile and flavour of foods [8]. Lipid oxidation, in particular, is a primary consideration for the meat, fish and dairy industries as it affects the sensory attributes and shelf-life of produce [9].

Antioxidants neutralise reactive oxygen and reactive nitrogen species (ROS/RNS, respectively), which can cause oxidative stress and damage to cells and tissues and disrupt essential functional physiological processes, including energy metabolism and the biosynthesis of proteins and peptides [10]. As an antioxidant, H_2 is demonstrated to quench cytotoxic free radicals, enhance endogenous antioxidant capacity and modulate both inflammatory responses (in animals) [11,12] and stress responses (in animals and in plants) [11,13]. Oxidation of biomolecules can also be inhibited with the post-harvest treatment of plant products (e.g., exposure to H_2 gas or treatment with hydrogen-rich water (HRW)). Additionally, exposing dairy products such as butter to H_2 during processing is shown to prevent biogenic amine formation due to microbial spoilage, a phenomenon attributed to the reducing effect of H_2 [14]. The ability of H_2 to traverse biological boundaries is likely to be highly influential, affecting fundamental organelle biochemistry as well as cytosolic reactions, a possible explanation as to why the application of H_2 has been demonstrated to have positive effects in numerous disease models in both plants and animals.

When growing plants or rearing animals, H_2 can be used as an additive to water in the form of HRW or hydrogen-rich nanobubble water (HNW), or as a food or soil supplement in the form of hydrogen-forming magnesium (Mg) or calcium (Ca) powders. Additionally, HRW, HNW and H_2 gas can also be used as a topical preparation on fruits and foliage (Figure 1).

As a gas, H_2 can be used during the dehydration processes of fruits and pulses, for example, and as hydrogen-infused packaging films in modified packaging atmospheres (Figure 1). Such processes are likely to reduce the oxidation of produce and prolong the shelf-life of foods [10]. Furthermore, foodborne pathogens also pose a significant risk to food security; here, H_2 is documented to inhibit the growth of such bacteria as *Escherichia coli, Salmonella listeriosis* and moulds [15], potentially reducing the risk of spoilage and foodborne illnesses.

Figure 1. Potential methods of utilizing hydrogen in the agri-food industry. Images created by the author using an artificial intelligence platform (wonderai.app). Mode: Photo. Prompts: HRW/HNW Pre-Harvest Irrigation 'Rice field irrigation'. Soil supplement 'Adding calcium powder to soil'. Post-Harvest 'Washing apples on farm'. H_2 Gas. Drying 'Drying fruit industry'. Packaging 'Dry packaging of food for retail'.

2. H_2 Use on the Farm

Global losses of food productivity as a result of increasing burdens (e.g., climate change, pollution and soil erosion) are estimated to be in the region of USD 170 billion annually [16]. Ergo, there is a pressing requirement for additional, effective, means of preserving the nutritional content, preventing the spoilage and prolonging the shelf-life of fresh produce. In this regard, there is an extensive body of research attesting to the benefits of adding H_2 into the feedwater of seeds, seedlings and maturing plants, wherein H_2, supplied as HRW, is demonstrated to improve crop viability and yield [17,18]. The antioxidant activity of H_2 is proposed to be the primary mode of action, involving the upregulation of glutathione and sulphur metabolism [19], enhanced antioxidant gene expression [18] and promoting hormone synthesis [20]. Such cytoprotective actions proffer resistance to both biotic and abiotic stressors, including drought [21], heavy metal toxicity [17,22], increased salinity [23], temperature variation [18,24] and viral infection [25].

A wealth of academic studies show that exposing plants to H_2 during the germination, plantation and growth phases can enhance root establishment, increase disease resistance and promote the growth and development of leaves, fruits and seeds. Incorporating H_2 into the feedwater of plant crops is a relatively simple process which can be implemented in various ways. For example, HRW and H_2 gas can also be used as a topical application. However, atmospheric losses due to evaporation make it difficult to envisage how practical this application method would be at scale. HRW and HNW can also be obtained by diffusing H_2 gas (formed through water electrolysis) into nutrient solutions and irrigation systems and by dissolving Mg or H_2 donors (e.g., ammonia borane hollow mesoporous silica nanoparticles (AB@hMSN)) in aqueous solutions. Such methods are already being implemented academically [18,20,26] and are beginning to be rolled out commercially.

The addition of H_2 into soil environments is also noted to improve soil quality through increasing H_2-oxidising microbial activity and promoting carbon fixation [27–29], a term referred to as hydrogen fertilisation, although the process is somewhat complex, which may rule out its application in an agricultural setting. Whether this type of application would provide a large-scale, long-term solution to replenishing nutrient-deficient soils will need to be explored.

3. H_2 in the Storage and Distribution Chain

The distribution and storage of fresh produce are paramount to ensuring food security with many aspects (e.g., transportation, warehouse and vendor) reliant on controlling the temperature at which foods are stored. The refrigeration process is not only energy-intensive, costly and environmentally taxing [30], but using this preservation method does not prevent, but mitigates, losses of perishable foods. Therefore, to ensure future food security, it is pertinent to consider novel approaches that can further reduce the deterioration rate of valuable produce.

A broad range of research in animals [11,31,32] and plants [33–35] identifies H_2 as having noticeable reducing properties, which are relevant to the preservation of food products as such redox poise favours a reducing environment, attenuating the accumulative oxidation of lipid components. Lipid oxidation is instrumental in the deterioration of biological substances as it can diminish cell membrane integrity, resulting in ion leakage, dysfunctional organelle activity and aberrant cellular cohesion [36,37]. Preventing lipid oxidation is a leading consideration for increasing the longevity and nutritional value of food products [9,10].

To date, research into H_2 describes the molecule as an effective mediator of cellular homeostasis, albeit the precise mode of action has yet to be delineated. Many reports highlight a direct scavenging effect of such ROS/RNS as the hydroxyl radical ($^{\bullet}OH$) and peroxynitrite ion ($ONOO^-$) [38–40], both significant contributors to lipid oxidation. However, considering the spatial and temporal distribution, and the kinetics of these highly reactive molecules, whether such reactions occur in vivo is questioned [41,42]. Alternative mechanisms of action involving the partial reduction in Fe^{3+} moieties [43] and protein stabilisation [44] have been proposed, and whilst the modality of H_2 is currently a matter of academic debate, what is clear is that implementation of HRW and H_2 gas to post-harvest produce consistently demonstrates an increased resistance to ripening, senescence and spoilage in fruits, herbs and vegetables (Table 1).

Table 1. Identifies the benefits of H_2 application to animal and plant products, noting the mechanism of action and overall effects on fresh produce. APx—ascorbate peroxidase; β-Gal—β-galactosidase; CAT—catalase; Cx—connexin; GR—glutathione reductase; GSH—glutathione; PG—peptidoglycan hydrolase; PL—pectate lyase; PME—pectin methylesterase; POD—peroxidase; SOD—superoxide dismutase.

Product	Application	Process	Mechanism	Outcome	Ref(s)
Banana (*Musa* spp. *AAA cv. Baxijiao*)	HRW (0.4 mM)	26-day storage	Reduced ethylene production and respiratory rate	Delayed ripening	[45]
Chive (*Allium tuberosum*)	H_2 gas (3%)	8-day storage	Increased antioxidant activity APx, CAT, GSH, POD and SOD	Increased shelf-life (100%)	[34]
Kiwi (*Actinidia chinensis*)	HRW (0.6 mM)	16-day storage	Increased antioxidant capacity (SOD) and reduced cellulase, PG and PME activity	Reduced respiration and rotting	[46]
Kiwi (*Actinidia chinensis*)	HRW (0.15 mM)	10-day storage	Increased antioxidant capacity APx, CAT, POD and SOD; reduced activity of β-Gal, Cx, PG, PL and PME	Delayed ripening	[47]
Mushrooms (*Hypsizygus marmoreus*)	HRW (0.1–0.4 mM)	12-day storage	Activated antioxidant activity (APx, CAT, GR, SOD); increased the contents of polysaccharides, protein, total sugar and amino acids	Extended shelf-life	[22,48]
Okra (*Abelmoschus esculentus* L.)	HRW (0.22 mM)	15-day storage	Increased biosynthesis of pectin, hemicellulose and cellulose	Delayed ripening	[49]
Pak Choi (*Brassica Chinensis*)	HRW (0.21 mM)	4-day storage	Delayed chlorophyll degradation; preserved phenolic content	Delayed senescence	[50]
Tomato (*Solanum lycopersicum*)	HRW (0.2 and 0.6 mM)	16-day storage	Prevented nitrite accumulation	Increased vitamin C levels	[51]

In addition to postharvest treatments, preharvest irrigation with HRW has also been demonstrated to protect natural produce against storage-related chilling injury [52]. This indicates that the application of H_2 has long-lasting, positive effects on plant biochemistry. As is shown in Table 1, H_2 has pleiotropic effects in plants, from regulating the activity of endogenous antioxidants including ascorbate peroxidase (APx), catalase (CAT), peroxidase (POD) and superoxide dismutase (SOD) [34], to reducing biomarkers of senescence β-Galactosidase (β-Gal), peptidoglycan hydrolase (PG) and pectin methylesterase (PME) [47] and increasing the biosynthesis of such polysaccharides as cellulose and pectin [49]. It is conceivable that, together, these cytoprotective qualities can further prevent the deterioration of fresh produce in storage and transit.

Incorporation of H_2-producing technologies into established storage and distribution chains will require careful consideration due, in part, to the possibility of igniting H_2 [53], a highly explosive gas, and to the time involved in washing and subsequently drying produce. Additionally, although the preparation of HRW baths is a relatively simplistic process [45,50], with a large volume of produce the process could become labour-intensive. Therefore, it may be prudent to focus on the pre-harvest and processing/packaging applications of H_2.

4. H_2 in Food Processing

A more practical and viable way of treating plant and animal produce can be envisaged during the packaging and processing stages, where H_2 can be applied without much adaptation to current preservation methods. H_2 gas can be easily incorporated into numerous preservation techniques, including drying, fermentation, extraction and preparation of foods (Table 2).

Table 2. The benefits of H_2 addition in the processing of food products. Ca—calcium; K—potassium; Mg—magnesium; Na—sodium. Ile—isoleucine; Leu—leucine; Met—methionine.

Product	Application	Process	Mechanism	Outcome	Ref(s)
Apricot (*Prunus armeniaca* L.)	H_2 gas (4%)	Drying	Increased antioxidant activity	Maintained nutritional profile	[54]
Beetroot (*Beta vulgaris*)	H_2-infused brine (undisclosed concentration)	Fermentation	Increased fermentation-associated microbial content	Reduced biogenic amine formation	[55]
Beetroot (*Beta vulgaris*)	H_2-enriched solvents (ethanol, methanol, water)	Nutrient extraction	Increased extraction yields: antioxidants, flavonoids and phenolics	Sustainable and efficient extraction process	[56]
Lemon (*Citrus lemon* L.)	H_2-enriched solvents (ethanol, methanol, water)	Nutrient extraction	Increased extraction yields: flavonoid and phenolic	Sustainable and efficient extraction process	[57]
Raw butter	HRW 1.6 mg/L	Washing during preparation	Reduced microbial decarboxylase activity; reduced biogenic amine formation; reduced yeast/mould	Improved shelf-life and sensory attributes. Reduced potential for foodborne toxicity	[14,58]
Rice milk	HRW (0.8 mM)	Preparation	Increased essential mineral content (Ca, K, Mg, Na); increased amino acids (Ile, Leu, Met)	Enhanced nutritional profile	[59]
Yoghurt	H_2 gas infusion into pasteurised milk (1 L/min–10 min)	Fermentation	Increased the acidification capacity: *Streptococcus thermophilus* (+10%); increased the reducing capacity: *Lactobacillus delbrueckii* (+13.7%)	Increased microbial fermentation activity	[60]

Dairy products are particularly susceptible to oxidation-related spoilage which can shorten the shelf-life. For example, oxidative reactions in butter lead to the destruction of essential fatty acids and vitamins. The foremost oxidative decomposition products formed in animal-derived produce such as butter are unstable hydroperoxides (R-OOH), molecules which are further decomposed into rancidity-inducing compounds (e.g., aldehydes (R-CH=O), ketones (R-C=O), alcohols (R-C-OH)) [61]. Introducing H_2 in the form of HRW when washing raw butter [58,60] was demonstrated to improve the quality attributes (e.g., acidity, colour), decrease the accumulation of heavy metals, prevent spoilage by inhibiting microbial decarboxylase activity, and retard the growth of yeasts and moulds, prolonging the shelf-life.

Adding H_2 into the preparation stages of such dairy products as butter and yoghurt is shown to inhibit microbial spoilage whilst also promoting microbial fermentation. Similarly, dissolving H_2 gas into the pickling media of vegetable produce was noted to enhance the fermentation process and reduce lactic-acid-bacteria-associated biogenic amine formation. The authors demonstrated that using HRW during the fermentation phase and replacing the medium with freshly infused H_2-enriched brine before pasteurisation was most effective in reducing levels of such biogenic amines as 2-phenylethylamine, putrescine and tryptamine [55]. Dissolving H_2 into solvent fluids used for the extraction of phytochemicals can increase the yields of anthocyanins, flavonoids and polyphenolic substances, whilst in the preparation of rice milk, an alternative to dairy, replacing pure water with HRW increased essential mineral and amino acid content [59].

The use of 4% H_2 gas in the fruit-drying process has also shown an increased potential to maintain the nutritional and sensorial notes of dried fruits [54]. Here, instead of infusing liquids with H_2 generated on demand, a mixture of gases (typically, CO_2, N_2 and H_2) into the drying apparatus is regulated via instrumentation. As such apparatuses utilise gas canisters, and since H_2 is explosive above concentrations of 75% (*v*/*v*) [62], increased storage and safety protocols may be required if this mechanism of application is to be exploited in the food preservation industry.

5. Incorporating H_2 in the Packaging Environment of Food

Historically, the packaging of food had the sole intention of providing a physical barrier between the environment and the contents to prevent external contamination. Innovation in the food packaging industry is broadly focused on the incorporation of active components such as ethylene scavengers (e.g., potassium permanganate) into the package, which can maintain, or extend, product quality, sensory attributes and shelf-life [63].

Currently, to improve the quality and longevity of produce, many fresh products are distributed and displayed in plastic packages containing a modified atmosphere composed of carbon dioxide (CO_2) and Nitrogen (N_2) [9]. This O_2-depleted environment inhibits cellular oxidative mechanisms and microbial activity, restricting senescent processes and preserving the contents. However, with compounding stresses on food security and distribution networks, any improvements which can enhance the longevity of fresh produce will be highly valued.

As with the utilisation of H_2 gas in thermal drying discussed previously, incorporating H_2 into the packaging atmosphere should be relatively simple. Nevertheless, unless the gas is used as it is produced, through water electrolysis, for example, the same safety concerns over the storage of pressurised gas and flammability apply. Therefore, stringent safety procedures in food processing and manufacturing environments will be necessary. On the other hand, the benefits of introducing non-explosive (<4% *v*/*v*) amounts of H_2 into packaging atmospheres may well outweigh any required investment in safety measures.

Although the precise mechanisms behind H_2 activity in cells have yet to be elucidated, the ability of H_2 to act as an effective antioxidant in plants and produce is now well-documented [8,64–66]. One of the primary functions of H_2 utilization in both agricultural and clinical settings is as an antioxidant, first demonstrated by Ohsawa et al. (2007), in a rodent model of ischemia/reperfusion injury [38]. The highly influential study describes the selective reduction in the highly reactive $^{\bullet}OH$ radical and $ONOO^-$ molecules, but no reduction in the important signalling molecules hydrogen peroxide (H_2O_2) and nitric oxide ($NO^{\bullet}$). Either as a result of, or in addition to, selectively neutralising the most detrimental reactive oxygen/nitrogen species, the upregulation and increased expression of such antioxidant proteins and peptides CAT, nuclear erythroid factor-2 (Nrf-2) and SOD are often reported [34]. Upregulation of such endogenous antioxidants is known to enhance tolerance to such environmental stressors as diseases, salinity, drought and temperature stress, and to confer a level of microbial resistance [23–25]. Such antioxidant activity is also demonstrated when H_2 is incorporated into the packaging atmosphere. Recently, several studies have identified that adding up to 4% H_2 gas into such packaging environments can markedly improve both the sensory qualities and shelf-life of fresh produce (Table 3).

Table 3. Identifies the benefits of H_2 application in the processing and packaging of fresh produce.

Product	Application	Mechanism	Outcome	Ref(s)
Cheese	H_2 gas (4%)	Reduced mesophilic-aerobic bacteria; reduced yeast-mould	Improved shelf-life	[67]
Eggs	H_2 gas (up to 3%)	Inhibited microbial spoilage; maintained antioxidant capacity; reduced pH (yolk and white)	Improved shelf-life	[68]
Freshwater Fish: Rainbow trout (*Oncorhynchus mykiss*)	H_2 gas (4%)	Reduced microbial decarboxylase activity; reduced biogenic amine formation; reduced protein degradation	Improved shelf-life Reduced potential for foodborne toxicity	[6]
Seawater Fish: Horse mackerel (*Trachurus trachurus*)	H_2 gas (4%)	Reduced microbial decarboxylase activity; reduced biogenic amine formation	Improved shelf-life; reduced potential for foodborne toxicity	[6]
Shrimp (*Fenneropenaeus chinensis*)	H_2 gas (up to 1%)	Reduced lipid oxidation and purine metabolism	Improved shelf-life and sensory attributes	[69]
Strawberries (variety undisclosed)	H_2 gas (4%)	Increased phenolic and anthocyanin content	Increased shelf-life (300–500%)	[70]

It can be seen from Table 3 that the addition of H_2 in the packaging atmosphere of animal and animal-derived products, including cheese, eggs and fish, was able to inhibit both lipid oxidation and the decomposition of amino acids by decarboxylase enzymes, whilst concomitantly reducing microbial growth and spoilage. As with the pre-harvest and processing applications of H_2 (Tables 1 and 2, respectively), the preservation of endogenous antioxidant activity may account for many of the effects noted in Table 3. Furthermore, in addition to being utilised as a preservative gas, H_2 gas can be used to detect flaws in food packages such as mis-sealing, tears, and small holes [71]. Here, the gas is used during the sealing process, before the packet passes through a sensor; if the sensor detects H_2, an alarm is activated, allowing for the damaged packet to be removed from the production chain and repackaged before wastage occurs.

Although supplementary research into the use of H_2 gas in the packaging environments of perishable goods is warranted, the initial findings strongly suggest that such reducing-atmosphere packaging environments can impede the natural deterioration of fresh produce. However, before this method of preservation can be adopted by the agri-food industry, detailed cost-benefit analyses, assessments of the practical incorporation into working spaces (e.g., safety, storage, transportation) and the effects of H_2 on other food products will be required.

6. Safety

Using H_2 safely will require proper handling precautions and procedures due to its flammability and potential for explosive reactions. For example, during storage, H_2 will need to be contained in appropriate cylinders in well-ventilated areas away from heat sources, ignition and direct sunlight [72], a factor that is likely to pose an issue towards comprehensive use, particularly in countries such as the UK and USA which have stringent safety policies. Cylinders should also be regularly checked for damage to reduce the potential of leakage. Additionally, as H_2 is lighter than air, proper airflow and ventilation are also necessary to prevent the accumulation of H_2. Equipment, fittings and connections of any apparatus using H_2 will also require detectors/monitors, which can provide early warning of malfunction and H_2 leakage. Pressure regulators and flow meters can detect sudden pressure changes and may well be instrumental in preventing harm from faults in H_2-associated equipment. It will be prudent to ensure individuals working with H_2 are adequately trained in the safe handling, storage and emergency procedures, with training to include recognizing the potential hazards and how to respond to incidences that may involve H_2 (e.g., evacuation, fires, leaks, etc.).

7. Future Perspectives and Conclusions

Reducing food waste and cutting costs is a primary concern for food security and the agri-food industry [73,74]. To better assess the potential for H_2 to support food growth, distribution and production industries, large-scale, in-field research, along with further empirical investigations into the primary, secondary and, perhaps, tertiary effects of H_2 applications will be advantageous. It may also be prudent to analyse whether heritable traits are affected by long-term H_2 usage. This could involve understanding the epigenetic context, hormone regulation, and resistance to biotic and abiotic stressors.

Recent studies, such as the one conducted by Cheng et al. (2021), have identified that HNW is suitable for irrigating crops [17]; therefore, it may be propitious to focus on this aspect of H_2 production when considering pre-harvest administration. For post-harvest applications, comparative analyses of whether treatment with HRW/HNW or H_2 gas, and at what stage in the distribution chain application is most effective, would be beneficial.

It is not only the efficacy of hydrogen treatments that must be scrutinised, but the cost-effectiveness, sustainability and long-term benefits of treatment must also be assessed. The cost of established 'green' hydrogen production technologies has been estimated to be 0.7–1.4 GBP/W for alkaline water electrolysis and 0.8–2.2 GBP/W for proton exchange membrane (PEM) electrolysis [75,76]. However, as such production costs are influenced by commercial innovation and investment, and the interest in H_2-producing technologies is growing internationally, it is likely such costs will be reduced as H_2 technology becomes more efficient over time. In addition to the purchase and maintenance costs, an assessment of the durability and longevity of commercial units versus conventional treatments and processes will also need to be considered.

To summarise both the risks and benefits of commercial H_2 usage in the agri-food industry, Table 4 provides an analysis of the strengths, weaknesses, opportunities and threats (SWOT) for incorporating H_2 into the agri-food chain.

Table 4. A SWOT analysis of strengths, weaknesses, opportunities and threats to incorporating H_2 into the agri-food chain.

	Strengths	Weaknesses	Opportunities	Threats
HRW/HNW				
Irrigation	Can be used in multiple stages of plant development (Germination, Growth) Low maintenance Measurable dosing	Retention of H_2	Agriculture Horticulture Floriculture	Unknown costs for large-scale implementation
Topical	Can be used pre- and post-harvest May reduce post-harvest spoilage	Possibly labour-intensive	Agriculture Horticulture Floriculture	Possibly difficult to implement Unknown commercial efficacy
H_2 Gas				
Drying	Can be used with a wide range of products Relatively easy to incorporate into established processes	Should not be used with high-temperature food preservation methods Should not be used near sources of ignition	Food manufacture and processing	Requirement for increased safety and storage protocols Unknown costs for large-scale implementation
Packaging	Can be used with a wide range of products Measurable dosing Relatively easy to incorporate into established processes	Should not be used near sources of ignition	Food packaging	Requirement for increased safety and storage protocols Unknown costs for large-scale implementation
Soil Supplement	Low maintenance	Unknown H_2 levels	Agriculture Horticulture Floriculture	Unknown long-term effects of bi-products on soil composition

In conclusion, via regulating antioxidant, hormonal and microbial activity, a wealth of evidence is accumulating strongly indicating that H_2 may be an effective fertiliser, microbial deterrent and preserving agent for edible produce (Tables 1–3). The application of H_2, whether as an atmospheric gas or in solution, into the agri-food industry may help to support the healthy growth of plant produce, increasing biomass, resistance to stress and yield (Table 1). Adopting the hydrogen strategy in various stages of the food production chain can decrease the heavy metal content, limit the biogenic amine formation of food products, preserve the nutritional and sensory properties, and extend the product's shelf-life [4,6,17,22]. Furthermore, reduction in both animal-derived and plant-derived food wastage can be achieved through H_2-induced inhibition of endogenous senescent processes (Tables 2 and 3). Moreover, as a sustainable, non-toxic and non-polluting agent, the future of H_2 application in the agri-food industry could help to reduce the carbon footprint, from the field to the table, of such a demanding supply chain.

Author Contributions: Conceptualization, G.R. and J.T.H.; writing—G.R.; writing—review and editing, G.R., J.T.H. and A.N. All authors have read and agreed to the published version of the manuscript.

Funding: This research was co-funded by Water Fuel Engineering and the University of the West of England. Funding identification number 7096050. Project code: RDAS0184.

Conflicts of Interest: Author Alexander Nenov was employed by the company Water Fuel Engineering. The remaining authors declare that the research was conducted in the absence of any commercial or financial relationships that could be construed as a potential conflict of interest.

Abbreviations

APx—Ascorbate peroxidase; β-Gal—β-Galactosidase; CAT—Catalase; Cx—Connexin; GR—Glutathione reductase; GSH—Glutathione; GRAS—Generally regarded as safe; HNW—Hydrogen-rich nanobubble water; HRS—Hydrogen-rich saline; HRW—Hydrogen-rich water; Nrf2—Nuclear erythroid factor-2; PEM—Proton exchange membrane; PG—Peptidoglycan hydrolase; PL—Pectate lyase; PME—Pectin methylesterase; POD—Peroxidase; RNS—Reactive nitrogen species; ROS—Reactive oxygen species; SOD—Superoxide dismutase; USD—United States dollar.

References

1. King, T.; Cole, M.; Farber, J.M.; Eisenbrand, G.; Zabaras, D.; Fox, E.M.; Hill, J.P. Food safety for food security: Relationship between global megatrends and developments in food safety. *Trends Food Sci. Technol.* **2017**, *68*, 160–175. [CrossRef]
2. Amit, S.K.; Uddin, M.; Rahman, R.; Islam, S.M.R.; Khan, M.S. A review on mechanisms and commercial aspects of food preservation and processing. *Agric. Food Secur.* **2017**, *6*, 51. [CrossRef]
3. Food and Drug Administration. GRASNotices. 2014. Available online: https://www.cfsanappsexternal.fda.gov/scripts/fdcc/?set=GRASNotices&id=520&sort=GRN_No&order=DESC&startrow=1&type=basic&search=hydrogen (accessed on 14 January 2024).
4. EU Commission Regulation. No 1129/2011 of 11 November 2011 Amending Annex II to Regulation (EC) No 1333/2008 of the European Parliament and of the Council by Establishing a Union List of Food Additives Text with EEA Relevance. Document Number: 32011R1129. Available online: https://eur-lex.europa.eu/legalcontent/EN/ALL/?uri=CELEX:32011R1129 (accessed on 14 July 2022).
5. Food Standards Agency. Approved Additives and E Numbers. 2018. Available online: https://www.food.gov.uk/businessguidance/approved-additives-and-e-numbers (accessed on 14 January 2024).
6. Singla, M.K.; Nijhawan, P.; Oberoi, A.S. Hydrogen fuel and fuel cell technology for cleaner future: A review. *Environ. Sci. Pollut. Res.* **2021**, *28*, 15607–15626. [CrossRef]
7. Ohta, S. Molecular hydrogen as a novel antioxidant: Overview of the advantages of hydrogen for medical applications. *Methods Enzymol.* **2015**, *555*, 289–317. [PubMed]
8. Alwazeer, D.; Çiğdem, A. Use of the molecular hydrogen in agriculture field. *Turk. J. Agri-Food Sci. Tech.* **2022**, *10*, 14–20.
9. Adegoke, G.O.; Kumar, M.V.; Krishna, A.G.; Varadaraj, M.C.; Sambaiah, K.; Lokesh, B.R. Antioxidants and lipid oxidation in foods-A critical appraisal. *J. Food Sci. Tech.* **1998**, *35*, 283–298.
10. Sezer, Y.; Bulut, M.; Boran, G.; Alwazeer, D. The effects of hydrogen incorporation in modified atmosphere packaging on the formation of biogenic amines in cold stored rainbow trout and horse mackerel. *J. Food Compos. Anal.* **2022**, *112*, 104688. [CrossRef]
11. Li, J.; Ge, Z.; Fan, L.; Wang, K. Protective effects of molecular hydrogen on steroid-induced osteonecrosis in rabbits via reducing oxidative stress and apoptosis. *BMC Musculoskelet. Disord.* **2017**, *18*, 58. [CrossRef]

12. Qi, D.-D.; Ding, M.-Y.; Wang, T.; Hayat, M.A.; Liu, T.; Zhang, J.-T. The therapeutic effects of oral intake of hydrogen rich water on cutaneous wound healing in dogs. *Vet. Sci.* **2021**, *8*, 264. [CrossRef]
13. Hancock, J.T. Editorial for Special Issue: "Production and Role of Molecular Hydrogen in Plants". *Plants* **2022**, *11*, 2047. [CrossRef]
14. Bulut, M.; Sezer, Y.Ç.; Ceylan, M.M.; Alwazeer, D.; Koyuncu, M. Hydrogen-rich water can reduce the formation of biogenic amines in butter. *Food Chem.* **2022**, *384*, 132613. [CrossRef] [PubMed]
15. George, S.M.; Richardson, L.C.; Pol, I.; Peck, M.W. Effect of oxygen concentration and redox potential on recovery of sublethally heat-damaged cells of Escherichia coli O157:H7, Salmonella enteritidis and Listeria monocytogenes. *J. Appl. Microbiol.* **1998**, *84*, 903–909. [CrossRef] [PubMed]
16. Razzaq, A.; Wani, S.H.; Saleem, F.; Yu, M.; Zhou, M.; Shabala, S. Rewilding crops for climate resilience: Economic analysis and *de novo* domestication strategies. *J. Exp. Bot.* **2021**, *72*, 6123–6139. [CrossRef] [PubMed]
17. Cheng, P.; Wang, J.; Zhao, Z.; Kong, L.; Lou, W.; Zhang, T.; Jing, D.; Yu, J.; Shu, Z.; Huang, L.; et al. Molecular hydrogen increases quantitative and qualitative traits of rice grain in field trials. *Plants* **2021**, *10*, 2331. [CrossRef] [PubMed]
18. Wang, X.; An, Z.; Liao, J.; Ran, N.; Zhu, Y.; Ren, S.; Meng, X.; Cui, N.; Yu, Y.; Fan, H. The role and mechanism of hydrogen-rich water in the *Cucumis sativus* response to chilling stress. *Int. J. Mol. Sci.* **2023**, *24*, 6702. [CrossRef]
19. Cui, W.; Yao, P.; Pan, J.; Dai, C.; Cao, H.; Chen, Z.; Zhang, S.; Xu, S.; Shen, W. Transcriptome analysis reveals insight into molecular hydrogen-induced cadmium tolerance in alfalfa: The prominent role of sulfur and (homo)glutathione metabolism. *BMC Plant Biol.* **2020**, *20*, 58. [CrossRef]
20. Felix, K.; Su, J.; Lu, R.; Zhao, G.; Cui, W.; Wang, R.; Mu, H.; Cui, J.; Shen, W. Hydrogen-induced tolerance against osmotic stress in alfalfa seedlings involves ABA signaling. *Plant Soil* **2019**, *445*, 409–423. [CrossRef]
21. Yan, M.; Yao, Y.; Mou, K.; Dan, Y.; Li, W.; Wang, C.; Liao, W. The involvement of abscisic acid in hydrogen gas-enhanced drought resistance in tomato seedlings. *Sci. Hortic.* **2021**, *292*, 110631. [CrossRef]
22. Zhang, J.; Hao, H.; Chen, M.; Wang, H.; Feng, Z.; Chen, H. Hydrogen-rich water alleviates the toxicities of different stresses to mycelial growth in Hypsizygus marmoreus. *AMB Express* **2017**, *7*, 107. [CrossRef]
23. Wu, Q.; Su, N.; Shabala, L.; Huang, L.; Yu, M.; Shabala, S. Understanding the mechanistic basis of ameliorating effects of hydrogen rich water on salinity tolerance in barley. *Environ. Exp. Bot.* **2020**, *177*, 104136. [CrossRef]
24. Guan, Q.; Ding, X.-W.; Jiang, R.; Ouyang, P.-L.; Gui, J.; Feng, L.; Yang, L.; Song, L.-H. Effects of hydrogen-rich water on the nutrient composition and antioxidative characteristics of sprouted black barley. *Food Chem.* **2019**, *299*, 125095. [CrossRef] [PubMed]
25. Shao, Y.; Lin, F.; Wang, Y.; Cheng, P.; Lou, W.; Wang, Z.; Liu, Z.; Chen, D.; Guo, W.; Lan, Y.; et al. Molecular Hydrogen Confers Resistance to Rice Stripe Virus. *Microbiol. Spectr.* **2023**, *11*, e0441722. [CrossRef] [PubMed]
26. Wang, Y.; Lv, P.; Kong, L.; Shen, W.; He, Q. Nanomaterial-mediated sustainable hydrogen supply induces lateral root formation via nitrate reductase-dependent nitric oxide. *Chem. Eng. J.* **2020**, *405*, 126905. [CrossRef]
27. Stein, S.; Selesi, D.; Schilling, R.; Pattis, I.; Schmid, M.; Hartmann, A. Microbial activity and bacterial composition of H2-treated soils with net CO2 fixation. *Soil Biol. Biochem.* **2005**, *37*, 1938–1945. [CrossRef]
28. Piché-Choquette, S.; Constant, P. Molecular hydrogen, a neglected key driver of soil biogeochemical processes. *Appl. Environ. Microbiol.* **2019**, *85*, 02418-18. [CrossRef] [PubMed]
29. Islam, Z.F.; Greening, C.; Hu, H.W. Microbial hydrogen cycling in agricultural systems–plant beneficial or detrimental? *Micro Biotech* **2023**, *16*, 1623–1628. [CrossRef] [PubMed]
30. Adekomaya, O.; Jamiru, T.; Sadiku, R.; Huan, Z. Sustaining the shelf life of fresh food in cold chain–A burden on the environment. *Alex. Eng. J.* **2016**, *55*, 1359–1365. [CrossRef]
31. Diao, M.; Zhang, S.; Wu, L.; Huan, L.; Huang, F.; Cui, Y.; Lin, Z. Hydrogen gas inhalation attenuates seawater instillation-induced acute lung injury via the Nrf2 pathway in rabbits. *Inflammation* **2016**, *39*, 2029–2039. [CrossRef] [PubMed]
32. Kalocayova, B.; Kura, B.; Vlkovicova, J.; Snurikova, D.; Vrbjar, N.; Frimmel, K.; Hudec, V.; Ondrusek, M.; Gasparovic, I.; Sramaty, R.; et al. Molecular hydrogen: Prospective treatment strategy of kidney damage after cardiac surgery. *Can. J. Physiol. Pharmacol.* **2023**, *101*, 502–508. [CrossRef]
33. Zeng, J.; Zhang, M.; Sun, X. Molecular hydrogen is involved in phytohormone signaling and stress responses in plants. *PLoS ONE* **2013**, *8*, e71038. [CrossRef]
34. Jiang, K.; Kuang, Y.; Feng, L.; Liu, Y.; Wang, S.; Du, H.; Shen, W. Molecular hydrogen maintains the storage quality of Chinese chive through improving antioxidant capacity. *Plants* **2021**, *10*, 1095. [CrossRef] [PubMed]
35. Tan, Y.; Duan, Y.; Chi, Q.; Wang, R.; Yin, Y.; Cui, D.; Li, S.; Wang, A.; Ma, R.; Li, B.; et al. The role of reactive oxygen species in plant response to radiation. *Int. J. Mol. Sci.* **2023**, *24*, 3346. [CrossRef] [PubMed]
36. Domínguez, R.; Pateiro, M.; Gagaoua, M.; Barba, F.J.; Zhang, W.; Lorenzo, J.M. A comprehensive review on lipid oxidation in meat and meat products. *Antioxidants* **2019**, *8*, 429. [CrossRef] [PubMed]
37. Sachdev, S.; Ansari, S.A.; Ansari, M.I. ROS Production and function at plasma membrane and apoplast. In *Reactive Oxygen Species in Plants: The Right Balance*; Springer Nature: Singapore, 2023; pp. 125–142.
38. Ohsawa, I.; Ishikawa, M.; Takahashi, K.; Watanabe, M.; Nishimaki, K.; Yamagata, K.; Katsura, K.-I.; Katayama, Y.; Asoh, S.; Ohta, S. Hydrogen acts as a therapeutic antioxidant by selectively reducing cytotoxic oxygen radicals. *Nat. Med.* **2007**, *13*, 688–694. [CrossRef] [PubMed]

39. Hanaoka, T.; Kamimura, N.; Yokota, T.; Takai, S.; Ohta, S. Molecular hydrogen protects chondrocytes from oxidative stress and indirectly alters gene expressions through reducing peroxynitrite derived from nitric oxide. *Med. Gas Res.* **2011**, *1*, 18. [CrossRef] [PubMed]
40. Kong, L.; Chen, C.; Mou, F.; Feng, Y.; You, M.; Yin, Y.; Guan, J. Magnesium particles coated with mesoporous nanoshells as sustainable therapeutic-hydrogen suppliers to scavenge continuously generated hydroxyl radicals in long term. *Part. Part. Syst. Charact.* **2019**, *36*, 1800424. [CrossRef]
41. Penders, J.; Kissner, R.; Koppenol, W.H. ONOOH does not react with H2: Potential beneficial effects of H2 as an antioxidant by selective reaction with hydroxyl radicals and peroxynitrite. *Free Radic. Biol. Med.* **2014**, *75*, 191–194. [CrossRef] [PubMed]
42. Li, Q.; Xie, F.; Yi, Y.; Zhao, P.; Zhang, X.; Zhang, X.; Zhang, X.; Ma, X. Hydroxyl-radical scavenging activity of hydrogen does not significantly contribute to its biological function. *BioRxiv* **2021**. [CrossRef]
43. Hancock, J.T.; LeBaron, T.W.; Russell, G. Molecular hydrogen: Redox reactions and possible biological interactions. *React. Oxyg. Species* **2021**, *11*, 17–25. [CrossRef]
44. Hancock, J.T.; Russell, G.; Craig, T.J.; May, J.; Morse, H.R.; Stamler, J.S. Understanding Hydrogen: Lessons to Be Learned from Physical Interactions between the Inert Gases and the Globin Superfamily. *Oxygen* **2022**, *2*, 578–590. [CrossRef]
45. Yun, Z.; Gao, H.; Chen, X.; Duan, X.; Jiang, Y. The role of hydrogen water in delaying ripening of banana fruit during postharvest storage. *Food Chem.* **2022**, *373*, 131590. [CrossRef] [PubMed]
46. Hu, H.; Li, P.; Wang, Y.; Gu, R. Hydrogen-rich water delays postharvest ripening and senescence of kiwifruit. *Food Chem.* **2014**, *156*, 100–109. [CrossRef] [PubMed]
47. Sun, Y.; Qiu, W.; Fang, X.; Zhao, X.; Xu, X.; Li, W. Hydrogen-rich water treatment of fresh-cut kiwifruit with slightly acidic electrolytic water: Influence on antioxidant metabolism and cell wall stability. *Foods* **2023**, *12*, 426. [CrossRef] [PubMed]
48. Chen, H.; Zhang, J.; Hao, H.; Feng, Z.; Chen, M.; Wang, H.; Ye, M. Hydrogen-rich water increases postharvest quality by enhancing antioxidant capacity in Hypsizygus marmoreus. *AMB Express* **2017**, *7*, 221. [CrossRef] [PubMed]
49. Dong, W.; Shi, L.; Li, S.; Xu, F.; Yang, Z.; Cao, S. Hydrogen-rich water delays fruit softening and prolongs shelf life of postharvest okras. *Food Chem.* **2023**, *399*, 133997. [CrossRef] [PubMed]
50. An, R.; Luo, S.; Zhou, H.; Zhang, Y.; Zhang, L.; Hu, H.; Li, P. Effects of hydrogen-rich water combined with vacuum precooling on the senescence and antioxidant capacity of pakchoi (*Brassica rapa* subsp. Chinensis). *Sci. Hortic.* **2021**, *289*, 110469. [CrossRef]
51. Zhang, Y.; Zhao, G.; Cheng, P.; Yan, X.; Li, Y.; Cheng, D.; Wang, R.; Chen, J.; Shen, W. Nitrite accumulation during storage of tomato fruit as prevented by hydrogen gas. *Int. J. Food Prop.* **2019**, *22*, 1425–1438. [CrossRef]
52. Hu, H.; Li, P.; Shen, W. Preharvest application of hydrogen-rich water not only affects daylily bud yield but also contributes to the alleviation of bud browning. *Sci. Hortic.* **2021**, *287*, 110267. [CrossRef]
53. Yang, F.; Wang, T.; Deng, X.; Dang, J.; Huang, Z.; Hu, S.; Li, Y.; Ouyang, M. Review on hydrogen safety issues: Incident statistics, hydrogen diffusion, and detonation process. *Int. J. Hydrogen Energy* **2021**, *46*, 31467–31488. [CrossRef]
54. Alwazeer, D. Reducing atmosphere drying as a new technique for the preservation of the color of dried foods. *J. Inst. Sci. Tech.* **2018**, *8*, 125–131.
55. Alwazeer, D.; Bulut, M.; Çelebi, Y. Hydrogen-rich water can restrict the formation of biogenic amines in red beet pickles. *Fermentation* **2022**, *8*, 741. [CrossRef]
56. Alwazeer, D.; Elnasanelkasim, M.A.; Çiğdem, A.; Engin, T.; Kanmaz, H.; Hayaloglu, A.A.; Russell, G.; Hancock, J.T. Hydrogen incorporation into solvents can improve the extraction of phenolics, flavonoids, anthocyanins, and antioxidants: A case-study using red beetroot. *Ind. Crop. Prod.* **2023**, *202*, 117005. [CrossRef]
57. Alwazeer, D.; Elnasanelkasim, M.A.; Çiğdem, A.; Engin, T.; LeBaron, T.W. Incorporation of molecular hydrogen into solvents increases the extraction efficiency of phenolics, flavonoids, anthocyanins, and antioxidants: The case of lemon peels. *Front. Sustain. Food Syst.* **2023**, *7*, 1223027. [CrossRef]
58. Ceylan, M.M.; Bulut, M.; Alwazeer, D.; Koyuncu, M. Evaluation of the impact of hydrogen-rich water on the quality attribute notes of butter. *J. Dairy Res.* **2022**, *89*, 431–439. [CrossRef] [PubMed]
59. Zor, M.; Bulut, M.; Karagöz, S.G.; Çetintaş, Y.; Alwazeer, D. Use of Hydrogen-Rich water in rice milk preparation improves the nutritional and sensory properties of product. *Food Chem.* **2024**, *437*, 137821. [CrossRef] [PubMed]
60. Bulut, M.; Alwazeer, D.; Tunçtürk, Y. Effects of the incorporation of hydrogen and nitrogen into milk on the reducing and acidification capacities of yoghurt bacteria. *J. Food Process. Preserv.* **2023**, *2023*, 7462909. [CrossRef]
61. Zamuz, S.; Bohrer, B.M.; Campagnol, P.C.B.; Domínguez, R.; Pateiro, M.; Santos, E.M.; Lorenzo, J.M. Lipid oxidation of animal fat. In *Food Lipids*; Academic Press: Cambridge, MA, USA, 2022; pp. 89–103.
62. Luo, J.; Liu, S.; Chen, P.; Lu, S.; Zhang, Q.; Chen, Y.; Du, B.; Tang, J.; He, J.; Liao, C.; et al. Fiber optic hydrogen sensor based on a Fabry–Perot interferometer with a fiber Bragg grating and a nanofilm. *Lab A Chip* **2021**, *21*, 1752–1758. [CrossRef]
63. Drago, E.; Campardelli, R.; Pettinato, M.; Perego, P. Innovations in smart packaging concepts for food: An extensive review. *Foods* **2020**, *9*, 1628. [CrossRef]
64. Zhang, X.; Zhao, X.; Wang, Z.; Shen, W.; Xu, X. Protective effects of hydrogen-rich water on the photosynthetic apparatus of maize seedlings (*Zea mays* L.) as a result of an increase in antioxidant enzyme activities under high light stress. *Plant Growth Regul.* **2015**, *77*, 43–56. [CrossRef]
65. Dai, C.; Cui, W.; Pan, J.; Xie, Y.; Wang, J.; Shen, W. Proteomic analysis provides insights into the molecular bases of hydrogen gas-induced cadmium resistance in Medicago sativa. *J. Proteom.* **2017**, *152*, 109–120. [CrossRef]

66. Liu, S.; Zha, Z.; Chen, S.; Tang, R.; Zhao, Y.; Lin, Q.; Duan, Y.; Wang, K. Hydrogen-rich water alleviates chilling injury-induced lignification of kiwifruit by inhibiting peroxidase activity and improving antioxidant system. *J. Sci. Food Agric.* **2022**, *103*, 2675–2680. [CrossRef]
67. Alwazeer, D.; Tan, K.; Örs, B. Reducing atmosphere packaging as a novel alternative technique for extending shelf life of fresh cheese. *J. Food Sci. Technol.* **2020**, *57*, 3013–3023. [CrossRef] [PubMed]
68. Wang, Y.; Wang, J.; Kuang, Y.; Shen, W. Packaging with hydrogen gas modified atmosphere can extend chicken egg storage. *J. Sci. Food Agric.* **2021**, *102*, 976–983. [CrossRef] [PubMed]
69. Jiang, K.; Zhang, Y.; Cai, C.; Lin, W.; Li, L.; Shen, W. Hydrogen-based modified atmosphere packaging delays the deterioration of dried shrimp (*Fenneropenaeus chinensis*) during accelerated storage. *Food Control.* **2023**, *152*, 109897. [CrossRef]
70. Alwazeer, D.; Özkan, N. Incorporation of hydrogen into the packaging atmosphere protects the nutritional, textural and sensorial freshness notes of strawberries and extends shelf life. *J. Food Sci. Technol.* **2022**, *59*, 3951–3964. [CrossRef] [PubMed]
71. Han, J.; Ruiz-Garcia, L.; Qian, J.; Yang, X. Food packaging: A comprehensive review and future trends. *Compr. Rev. Food Sci. Food Saf.* **2018**, *17*, 860–877. [CrossRef]
72. Abohamzeh, E.; Salehi, F.; Sheikholeslami, M.; Abbassi, R.; Khan, F. Review of hydrogen safety during storage, transmission, and applications processes. *J. Loss Prev. Process Ind.* **2021**, *72*, 104569. [CrossRef]
73. Britz, W.; Dudu, H.; Fusacchia, I.; Jafari, Y.; Roson, R.; Salvatici, L.; Sartori, M. *Economy-Wide Analysis of Food Waste Reductions and Related Costs*; Publications Office of the European Union: Luxembourg, 2019.
74. Bajželj, B.; Quested, T.E.; Röös, E.; Swannell, R.P. The role of reducing food waste for resilient food systems. *Ecosyst. Serv.* **2020**, *45*, 101140. [CrossRef]
75. Younas, M.; Shafique, S.; Hafeez, A.; Javed, F.; Rehman, F. An overview of hydrogen production: Current status, potential, and challenges. *Fuel* **2022**, *316*, 123317. [CrossRef]
76. Finke, C.E.; Leandri, H.F.; Karumb, E.T.; Zheng, D.; Hoffmann, M.R.; Fromer, N.A. Economically advantageous pathways for reducing greenhouse gas emissions from industrial hydrogen under common, current economic conditions. *Energy Environ. Sci.* **2021**, *14*, 1517–1529. [CrossRef]

MDPI AG
Grosspeteranlage 5
4052 Basel
Switzerland
Tel.: +41 61 683 77 34

Applied Sciences Editorial Office
E-mail: applsci@mdpi.com
www.mdpi.com/journal/applsci

www.ingramcontent.com/pod-product-compliance
Lightning Source LLC
Chambersburg PA
CBHW062100220926
51850CB00010B/2833